Desert Water

Desert Water

The Future of Utah's Water Resources

edited by

Hal Crimmel

THE UNIVERSITY OF UTAH PRESS
Salt Lake City

The Defiance House Man colophon is a registered trademark of the University of Utah Press. It is based on a four-foot-tall Ancient Pueblo pictograph (late PIII) near Glen Canyon, Utah.

18 17 16 15 14 1 2 3 4 5

LIBRARY OF CONGRESS CATALOGING-IN-PUBLICATION DATA

Desert water : the future of Utah's water resources / edited by Hal Crimmel.

pages cm

Includes bibliographical references and index.

ISBN 978-1-60781-375-0 (pbk. : alk. paper)

ISBN 978-1-60781-374-3 (ebook)

1. Water resources development—Utah. 2. Water-supply—Utah. I. Crimmel, Hal, 1966- editor of compilation.

HD1694.U8D47 2014

333.91009792—dc23

2014018323

Index by Andrew L. Christenson.

Printed and bound by Sheridan Books, Inc., Ann Arbor, Michigan.

Contents

Figures

Acknowledgments

This book is a deeply collaborative effort. I want to thank the authors who contributed their time and energy to the dual goals of informing readers of the many water issues that Utah faces and posing solutions to these problems. Thank you all for your passion, patience, and perseverance. I also want to thank Dr. John Alley, editor in chief at the University of Utah Press, for his encouragement and guidance from proposal to published book, and Phoebe McNeally at the DIGIT Lab at the University of Utah, for making the map. At Weber State, I want to thank my many colleagues who have worked with me over the years on environmental issues and helped me think more deeply about water issues: Jenn Bodine, Jake Cain, Larry Clarkson, Sara Dant, Bryan Dorsey, Eric Ewert, Therese Grijalva, Kevin Hansen, Shaun Hansen, Mike Hernandez, Susie Hulet, Kathryn Lindquist, Madonne Miner, Alice Mulder, John Mull, Mark Stevenson, Norm Tarbox, Ryan Thomas, Michael Vaughan, Mikel Vause, Jan Winniford, and many others. My assistant in the Master of Arts English program office, Genevieve Bates, has helped me in countless ways with this project. I also want to thank my wonderful immediate and extended family for their support on this and other projects; the children especially remind me why we need to think carefully about the future of water in Utah. Finally, I would like to thank all of those individuals and organizations working to ensure that our water resources contribute equally to a healthy environment, a healthy economy, and healthy people.

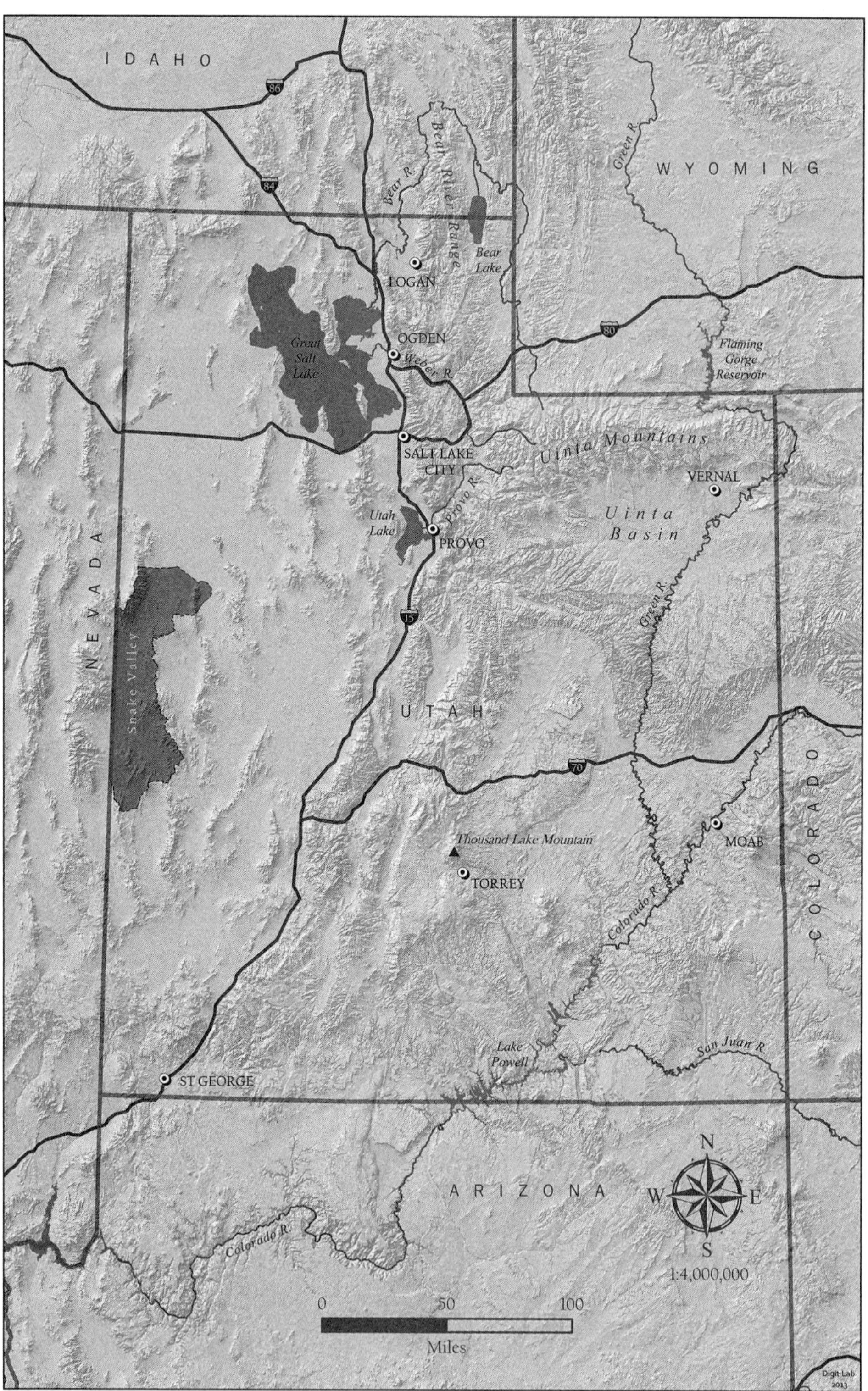

Figure 0.1. Map of Utah. DIGIT Lab, University of Utah.

Introduction

HAL CRIMMEL

THIS COLLECTION FROM leading scientists, writers, and policy makers explores the future of water in one of the nation's most rapidly growing states. That future is not far off in the distance—the future of Utah's water is here. Today's decisions will shape the amount and quality of water available to meet the future demands of a population expected to double by 2060 to nearly 6 million, based on projections from the Governor's Office of Management and Budget.[1] The specific goal of this book is to increase awareness across a broad segment of the population, including citizens, policy makers, students and teachers, water professionals, and others interested in expanding their knowledge of water-related issues in Utah. Rivers, creeks, and aquifers cross state lines, so Utah is not an island unto itself; decisions about water necessarily must take place within a broader regional context that includes neighboring states. In several chapters this book touches on how neighboring states affect Utah's water and how Utah's water decisions impact other states,[2] but the primary focus is on water resources within the state itself. Readers from other states, however, may find that the problems and solutions discussed here are relevant to their water resources as well, because many of the same climactic trends, water laws, policies, perceptions, and misconceptions are present elsewhere.

From the Bear River watershed and Great Salt Lake in the northern part of the state to the oil and gas fields of eastern Utah to the Snake Valley of western Utah to the Colorado River in the south, the collection explores the effects of climate change on mountain snowpack, the threats to groundwater, springs, streams, and rivers and the ecological and economic values of Great Salt Lake. The book also considers the challenges—primarily policy related but also associated with consumption practices—that need to be addressed for us to continue delivering water to growing rural and urban populations. Readers will also learn of issues that threaten Utahns' access to fishing and to floating our rivers and will come to a better understanding of water ethics and policies—some good, some bad—to identify those that might be changed. In addition to identifying problems, however, *Desert Water* proposes a number of workable solutions to these challenges by advocating for a fundamental shift in how we view water, by seeking to increase public understanding and support for forward-thinking policies, or by directly suggesting policy improvements.

All too often water availability and quality are taken for granted. Water enables human life, agriculture, and industry, yet it is also essential for a functioning ecosystem. For instance, "75% of the 360 bird species found in Utah depend on riparian areas for some part of their life cycle,"[3] as do other forms of wildlife. What use is a strong economy if our drinking water and the air we breathe are so toxic or misused that neither we nor the creatures that we share the world with can enjoy our day-to-day existence? China, with its terrible air and water pollution, illustrates the pitfalls associated with a form of economic development that gives short shrift to maintaining a healthy environment.

Those living in the American West can't help viewing water in a way that people in wetter regions of the country do not. In part this is because water is not visible on a daily basis as it is in the upper Midwest or New England, for instance. Citizens in those areas—if they fret about water at all—tend to worry about too much of it: widespread spring flooding from rivers swollen by snowmelt or from coastal hurricanes, though climate change is beginning to alter these ancient patterns.[4]

Although in the Intermountain West we occasionally see what seems to be an excess of water—think of flash floods in desert canyons, high water in Great Salt Lake in 1983, snow avalanches in the Wasatch or Uinta Mountains, or summertime cloudbursts that flood residential areas. But typically, and in Utah in particular, the main perception—fueled by a water lobby that prefers more development in the form of expensive dams, diversions, and pipelines over conservation—is that we are running out of water, a complicated notion that this books seeks to examine more closely.

Droughts have periodically swept the Intermountain West and especially Utah. Though drought is part of natural climatic cycles, during these dry periods our limited water resources are subjected to even more pressure than usual to provide for the needs of people, industry, and the environment. In Utah, as in other arid regions of the country, new water issues seem to emerge almost daily. In fact, it's hard to go more than a week without a new controversial water issue in the headlines. This pattern is only likely to continue, given the growing population. With no new conservation measures, for example, the state projects an 800,000 acre-foot annual water shortage by 2060,[5] though it is hoped that targeted conservation measures will reduce that deficit to 283,000 acre-feet.[6] Other issues pertain to water quality. For at least a decade now we've known that "fresh water supply, which is indispensable for both human well-being and environmental integrity, is on par with climate change as the two physical limitations most likely to change the way we live."[7] And 97 percent of Utahns indicate that it is "important to maintain water quality for future generations"; improving water quality in Utah's lakes, streams, rivers, and reservoirs would have a positive economic impact of over $1 billion.[8] Yet, according to the *Deseret News*, "Utah is staring at a $1.2 billion water pollution problem that state regulators say is the most severe issue they have confronted since the federal Clean Water Act was passed more than 40 years ago."[9] When a lake or stream in Utah exceeds pollution thresholds and is assessed as impaired, it is added to the "303(d)" list of impaired waters. According to a 2013 Utah Division of Water Quality report, 48 of 132 "priority" lakes and reservoirs are "not meeting their beneficial uses; all but 5 are 'listed' due to nutrient pollution,"[10] which comes from "nitrogen and phosphorous runoff from agriculture, homeowners, and industry."[11] Beneficial uses of surface waters include untreated water for culinary use, recreation, sustaining aquatic wildlife, and agriculture. Nutrient pollution negatively impacts all of these categories. But it is not just surface water that needs to be cleaned up. Statewide nineteen of Utah's public water systems have "elevated nitrate levels";[12] nitrates are a suspected human carcinogen and are also toxic to infants.

Other threats to the state's water resources include a proposed nuclear power plant near the town of Green River, Utah, that would require 53,000-plus acre-feet of Green River water for reactor cooling.[13] Some feel that the "nuclear promoters are speculating, without the means to build a plant, and may end up leasing the water for other energy projects in eastern Utah,"[14] in a possible bait-and-switch proposition. Either way, the impact on the Green River will be significant: energy projects create their own set of water issues, as oil and gas operations can use between 300,000 and 4 million gallons of water to drill and/or hydraulic fracture each well.[15] This water must be pumped from rivers or from wells and then be disposed of or recycled after the drilling (and/or fracking) process is complete. Companies currently drill 1,000 new wells in Utah each year,[16] with more than 25,000 new wells currently proposed for the Uinta Basin.[17]

Multiplying 1,000 new wells annually by the millions of gallons of water required to drill them makes it clear that this is not an insignificant amount of water. But pumping groundwater is not just an issue associated with oil and gas operations; groundwater pumping schemes proposed by the City of Las Vegas would dry out topsoil in vast tracts of valleys in the western part of Utah. That proposal—to pump millions of gallons of groundwater and send it through a 285-mile pipeline to Las Vegas—was not supported by Utah governor Gary Herbert in 2013. But no water district in the nation is more tenacious and persistent than the one headed by Las Vegas's Patricia Mulroy, and we have not heard the last of this issue. Another persistent threat is the pipeline proposal that would withdraw up to 250,000 acre-feet of Green River water—over 65 billion gallons annually—from Flaming Gorge Dam and pipe it to front-range Colorado cities. The developer continues to seek ways to gain approval, though so far his proposals have been rejected by a trio of federal agencies. Still, the threats to the Green River are so great that in 2012 American Rivers ranked the Green at number two on its "America's Most Endangered Rivers" list.

Other interests would use taxpayer money to build an expensive pipeline from Lake Powell that would benefit Washington and Kane Counties in southern Utah. This billion-dollar pipeline would deliver water primarily to St. George, a city that shows little propensity to conserve water. This proposed diversion of the Colorado River was one of the many reasons why American Rivers listed the Colorado as the most endangered river in the United States in 2013.[18] There are suggestions to add more expensive dams that would dewater tributaries such as the Bear River, which keep Great Salt Lake at levels that are economically and environmentally sustainable. The list is long, and no end to our water issues appears to be in sight. Unrelenting demands on the state's water resources will continue to be made and in light of our inefficient use, water will need to be managed better to secure the economic and environmental future of a growing state located in an arid region with a high per-capita rate of water consumption.

If these challenges are not enough, climate changes will add even more unpredictability and pressure on resources. Modeling done by atmospheric scientists using three different greenhouse gas scenarios to 2100 projects much warmer maximum air temperatures in the Wasatch Mountains. In March and April, for instance, the rise is projected to be as much as 50 percent—from 8 degrees Celsius to 12 degrees Celsius—meaning earlier and more intense spring runoff and also a diminished April snowpack, perhaps by more than 50 percent. Climate modeling also predicts a much drier southern half of the state and a wetter northern half, with most of that additional precipitation falling as rain.[19] Current water storage systems in the Wasatch are designed to capture snowmelt, so the additional precipitation falling as rain will require new techniques to capture that water and make it usable.

This book arose from that nexus of concerns. How should water resources best be managed for Utah (and by extension other arid states) to continue to thrive? How much water does Utah have, and how much water does it need? Thoughtful and sustainable answers to these questions in the context of the triple bottom line—people, profit, planet—will be of the utmost importance in the decades to come.

Readers will find that each chapter in this collection reads a little differently. Some chapters contain more lyrical narratives that tend to intertwine the personal experiences of each author with a particular place and issue; others are grounded more in science and policy; still others combine science, policy, and personal experience. Readers will find in general that a policy or scientifically oriented chapter is followed by a more personal narrative before returning to another policy or scientifically oriented chapter. I hope that those using this book will embrace this variety of approaches and value the multiple perspectives. I also hope that readers will be engaged by each writer's strong concern for the issue and ultimately will be swayed by the research that, where present, underpins the discussion. Some readers may prefer to read the book from front to back; others may enjoy reading in the opposite direction or prefer to skip around from chapter to chapter. Because of these different options, certain ideas or concepts may be repeated across chapters, if the concept seems essential for maintaining the coherence of an individual chapter.

Debates about environment issues such as water have become increasingly polarized, though I believe most Utahns want similar things despite their differing political leanings. We need to work to find common ground. "Liberals," for instance, could more widely acknowledge that an industrial society providing technology, transportation, and modern medicine for its citizens does need to use the earth's resources responsibly to provide for an increasingly urban, technologically advanced society. "Conservatives" might work to put "conserve" back into the notion of what it means to be a conservative. Many will have a family member who reuses aluminum foil or spends a day canning surplus fruits and vegetables from the garden. These individuals embrace what we might call Depression-era or pioneer-era values: thrift, conservation, sacrifice. Few reading this today will not on some level embrace such qualities that defined our ancestors' values, regardless of place of origin. These values of conservation—particularly as they apply to water—must be more widely championed by all, if we are to have enough water in the decades to come.

At the same time, despite widespread acceptance of such pioneer-era values, calling them "environmental" often derails the conversation, oddly enough. I'm not sure if the word "environmental" is as broadly inflammatory as the term "desegregation" was for some in the Jim Crow South, but in some circles it certainly is. I suspect, though, that reexamination of what the term "environmental" means reveals that it should stand for the core concerns of all citizens in the state—a strong economy that *also* provides a healthy environment for children, families, and wildlife and that uses the best available technologies and practices to ensure clean air and clean water for all of us.

If responsibility to use our water resources judiciously should be shared by all, it bears asking how does water contribute to a healthy, sustainable economy? And what obligations do individuals, industry, and government have to maintain a healthy environment that keeps lakes, streams, groundwater, and rivers free of pollutants and in so doing provides habitat for plants, trees, animals, fish, or birds? What obligations do we have to steward our water resources for future generations?

This collection begins with geographer Eric Ewert's "The Coming Challenge: Population Growth and Water Decline," which identifies many of the pressures on water resources, particularly those in urban Utah. Ewert paints in broad strokes the challenges faced statewide and in the Wasatch Front in particular, including population growth, inefficient agricultural water use, and a decline in winter snowpack, to name but a few, and suggests that we need to make changes in our water use practices if we are to enjoy a sustainable future.

Chapter 2, "The Miracle at the End of the Line," by award-winning writer Stephen Trimble, takes a personal look at the miracle of cold, clean water flowing out of his home faucet in rural Utah. Where does this water come from? How does it start clean and stay clean? In rural communities facing growing pains, can this gift of water help preserve an agricultural way of life?

In chapter 3, "Bear River: Learning from a River That Closes Our Circle," Craig Denton, author of a beautifully written and photographed book on the Bear River, takes readers to a watershed in the top of the state.[20] We learn that as Great Salt Lake's major tributary the Bear River provides about 60 percent of the water to the lake, which supports vital aquatic and avian ecosystems. But the river is under increasing pressure: without greater awareness and better protection for the Bear, this important natural resource will be degraded, negatively impacting the quality of life in northern Utah.

Depending on who gets asked, religious beliefs in Utah either contribute to a careless attitude toward Utah's natural resources in general (and water in particular) or contribute to an ethos of stewardship that provides the foundation for a sustainable future. In chapter 4, "The Restoration of All Things: The Case of the Provo River Delta," Brigham Young University professor George Handley explores how beliefs held by members of the Church of Jesus Christ of Latter-day Saints specifically inform the restoration of the Provo River Delta and suggests that these beliefs should form the basis for Mormons to embrace a more robust commitment for preserving and restoring the environment in general.

In chapter 5 the book turns in detail to a resource that rarely gets the attention of Utahns: Great Salt Lake. Expanding on the discussion in chapter 3 on the connection between the Bear River inflows and the health of Great Salt Lake, in chapter 5 geographer Daniel Bedford explores "Climate

Change and the Future of Great Salt Lake." He examines how the effects of a warming climate—a fact that policy makers and water managers across the state are no longer debating but actively preparing for—may influence inflows and water levels in the lake in coming years. Without factoring the effects of climate change into lake management plans, notes Bedford, northern Utah risks bearing witness to a shrinking lake, with serious impacts on bird habitat, brine shrimp, and air quality.

In chapter 6, "Chicken Little's New Career: How Utah's Water Development Industry Sows False Fears and Misinformation," Zachary Frankel describes how the proposed $1 billion-plus Lake Powell Pipeline would burden all Utah taxpayers with a project that is not needed to meet the future water needs of residents in Washington and Kane Counties. Frankel, executive director of the Utah Rivers Council, goes on to explain that property tax structures in St. George, for example, amount to a hidden subsidy for water, making it seem so cheap—fifteen times less than the rates that citizens of well-watered Seattle pay—that people have little incentive to conserve. Phasing out water waste, notes Frankel, would provide more than enough water for St. George and surrounding communities. But the problem of mispriced water is not limited to St. George: it is present in the Wasatch Front, the Bear River, and the Price River watershed. This problem is one that the state itself needs to tackle in order to find fiscally conservative solutions to Utah's water needs.

Economic considerations also play a role in discussions of Great Salt Lake's future. In chapter 7, "Time to Rethink Policy: Ideas for Improving the Health of Great Salt Lake," attorney Rob Dubuc focuses on tensions between the estimated $1.32 billion that the lake contributes to the state's economy and the need to maintain a healthy ecosystem that complements and supports the lake's economic value. The health of that ecosystem is contingent upon maintaining a certain amount of water in the lake, so Dubuc asks if revising water law to guarantee a certain minimum lake level would solve the problem. Along the way, the chapter also explores the many threats to the lake's long-term future, which include mineral extraction, nutrient loading, pollution, and issues resulting from causeways and diking in the lake.

Fishers, river runners, birders, and others who enjoy the tranquillity and perspective that a river provides will find themselves drawn to chapter 8, "The Colorado: Archetypal River," by writer and former river guide Brooke Williams. This chapter explores how a specific beach on the Colorado River near Moab contains Williams's hopes and fears for the future of the river, which is threatened by drought and a changing climate. Williams wonders if the Colorado will continue to be able to provide for the needs of humans and nature. Will it still be able to offer those rare moments of insight and reflection that provide a balance between our inner and outer worlds?

Such experiences walking to and spending time along the banks of state's rivers are threatened, as environmental historian Sara Dant points out in chapter 9, "Going with the Flow: Navigating to Stream Access Consensus." Her chapter focuses on current laws restricting public recreational access to Utah's streams. Should private property rights trump the public's right to access what many see as a public resource for recreational purposes—fishing, floating, or wading? Dant explores the history of river use in Utah, particularly the Weber River, and discusses the access issues as considered by Utah courts.

In chapter 10, "The Return of Glen Canyon: The Beginning of a More Sustainable Future for the Colorado," we move to southern Utah and to Glen Canyon on the Colorado, perhaps better known by many as Lake Powell. Journalism professor Annette McGivney makes it clear that Lake Powell is in a precarious situation. Half-full in 2013, the lake finds itself suffering from drought, a warming climate that increases evaporation and contributes to further water loss, and siltation that is filling in the upper reaches of the reservoir. McGivney, author of a leading book on Glen Canyon, suggests that dropping the level of Lake Powell would conserve water and also offer recreational opportunities in Glen Canyon that would draw visitors from around the world.

In chapter 11, "Land of 20,000 Wells: Impacts on Water from Oil and Gas Development in Eastern Utah," editor Hal Crimmel discusses the multitude of water use and water disposal issues associated with oil and gas drilling in the Uinta Basin and the Green River watershed. The risks are many, yet the industry typically downplays them by insisting that its methods are safe. Given the number of existing wells and plans for thousands more, it is possible that Utah is creating a Superfund-like problem that will contaminate water supplies and the environment for generations to come and may affect many of the 30 million downstream users of the Colorado River Basin. What policies and regulations might help us avoid such a scenario?

Similar short-term thinking is also evident in the groundwater pumping scheme threatening the Snake Valley in western Utah, whereby Las Vegas would pump water out of aquifers and pipe it to Las Vegas. In chapter 12, "Moving Water," writer Jana Richman, the daughter of a rancher from Utah's West Desert, alerts readers to the disastrous impact of the Las Vegas pumping and pipeline project on wildlife, springs, wetlands, and ranchers' livelihoods. In a broader context, says Richman, our civilization needs to find a way to live within the limits imposed on it by nature—and pumping water out of an ecologically sensitive valley that has the potential to dry up and create disastrous air pollution for all the Utahns living downwind is not a step toward acknowledging those limits.

Daniel McCool concludes the book with his chapter "A New Water Ethic." McCool, a political scientist, reminds us: "We do not have a water crisis in Utah, we have a water *management* crisis. We have a crisis of innovation and a lack of vision." This chapter asserts that we need to embrace a new water ethic now that abandons old models of thinking about water, particularly those that place the needs of one individual before the needs of all stakeholders, from people to the ecosystems that we depend upon for survival.

Today and in coming years the citizens of Utah, as well as those residing in other states throughout the West, will need to pay greater attention to water issues that ultimately concern everyone. Towns, cities, aquifers, rivers, lakes, streams, and all of the people, wildlife, and agriculture that depend on sustainable approaches to the issues set forth in this book demand no less. My hope is that these chapters will increase public awareness of the importance of water in the state, which will lead to more innovative and progressive policies that will benefit current and future generations.

Notes

1. "2012 Baseline City Population Projections," Governor's Office of Management and Budget, State of Utah, http://www.governor.utah.gov/dea/projections.html.

2. See chapter 3 by Craig Denton, chapter 10 by Annette McGivney, chapter 11 by Hal Crimmel, and chapter 12 by Jana Richman herein.
3. "Growing Wild," State of Utah, Division of Natural Resources, Division of Wildlife Resources, 1995, http://wildlife.utah.gov/education/newsletters/95spring-gw.pdf.
4. See Charles Fishman's *The Big Thirst: The Secret Life and Turbulent Future of Water*, which discusses how Atlanta nearly ran out of water during a sustained drought.
5. An acre-foot of water equals 326,000 gallons, enough water to flood an acre of land one foot deep or support 1,200 Utahns for a day.
6. Todd Adams and Eric Millis, State of Utah, Department of Natural Resources, Utah Division of Water Resources, "Planning for Utah's Water Needs" (presentation, American Water Resources Conference, Utah Section, Salt Lake City, May 14, 2013); some believe that this observation is more a claim than a finding of fact, however, and hence open to debate.
7. Sandra B. Zellmer and Jessica Harder, "Unbundling Property in Water."
8. "Economic Benefits of Reducing Nutrient Pollution in Utah's Waters," State of Utah, Department of Environmental Quality, Utah Division of Water Quality, 2012, http://www.waterquality.utah.gov/nutrient/documents/UtahDWQ_NutrientBenefits_ExecSummary_Final.pdf.
9. Amy Joi O'Donoghue, "Utah Faces Polluted Water Woe."
10. In 2008 a Utah Division of Water Quality report noted that roughly 32 percent of all the surface acreage of Utah's lakes (149,760 acres) and 28 percent of the streams and rivers assessed (3,080 miles) did not support at least one "beneficial use."
11. Walt Baker, "Utah's Approach to Regulating Nutrients," Utah Division of Water Quality 2013 Report to Subcommittee of the Utah State Legislature.
12. Ibid.
13. The *New York Times* has suggested that nuclear power "could become a bypassed technology—like moon landings, Polaroid photos and cassette tapes." See Matthew L. Wald, "Atomic Power's Green Light or Red Flag: Delayed Project in Georgia May Be a Road Map for the Industry's Future," *New York Times*, June 12, 2013, B1/B6.
14. Brandon Loomis, "Protestors Gather to Thwart Green River Nuclear Plans."
15. For further discussions and sources of data, see chapter 11 by Hal Crimmel herein.
16. John Baza, State of Utah, Department of Natural Resources, Utah Division of Oil, Gas, and Mining, "Hydraulic Fracturing: Water Resource Impacts of This Oilfield Well Stimulation Technique" (presentation, American Water Resources Conference, Utah Section, Salt Lake City, May 14, 2013).

17. Michael D. Vanden Berg, Utah Geological Survey, "Saline Water Disposal Issues in the Uinta Basin, Utah" (presentation, American Water Resources Conference, Utah Section, Salt Lake City, May 14, 2013).
18. "Most Endangered Rivers for 2013," American Rivers, http://www.americanrivers.org/endangered-rivers/2013-report/.
19. Court Strong, "Future Precipitation and Snowpack along the Wasatch Range" (presentation, American Water Resources Conference, Utah Section, Salt Lake City, May 14, 2013).
20. See Craig Denton, *Bear River: Last Chance to Change Course.*

1

The Coming Challenge

Population Growth and Water Decline

ERIC C. EWERT

ANYONE SITTING STREAMSIDE and resting comfortably in lush grass, deep in the shade of a broad-leafed tree, may find it hard to worry much about water supplies in Utah. All along the Wasatch Front we can find such green sanctuaries and then look upward at the snowcapped mountains and forget altogether that we live in a desert. The feeling of abundance is contrived, though, and the view is a mirage. The Kentucky bluegrass and sycamore trees were imported from far wetter parts of the country. They only grow here because of an elaborate system of reservoirs, pipes, and sprinklers. The gleaming white snowpack on the peaks will decline rapidly, mainly due to climate change. And the gurgling streams rarely reach their destination, Great Salt Lake. Instead water once bound for the lake grows grass and trees and crops and people. Over the chorus of the creek, we can easily hear the constant hum of highway traffic, the thwack-thwack of a house builder's nail gun, and the screech of children at recess outside an overstuffed elementary school. It's far easier to tune out the clatter of humanity and revel in the verdant stream banks, but we no longer have that luxury in our desert oasis.

The state of Utah is speeding toward a water and population train wreck. Actually, it's more of a head-on collision. Going fast and gaining speed in one direction is rapid population growth. And headed in the other direction is a declining fresh water supply. Unless our relationship to water changes soon, the collision of these two dominant trends will pose real challenges for everyone and everything in the state of Utah, from individual households to cities, farms, industry, and particularly for the plants and animals that rely on adequate sources of fresh water. In 2012 more than 100,000 people moved to Utah, but the vast majority of the state's reservoirs remained well below capacity after yet another dry winter. Numerous trends document that the supply of water in Utah is dwindling just as demands for water continue to increase, mainly driven by rapid population growth. This inverse relationship can only lead to hardship if new residents expect to waste water at the same rate as existing residents and if we hope to have anything resembling a healthy ecosystem. It doesn't have to be this way.

In other places, where water is scarce and the list of uses and users is growing, water managers have enacted many policies and commonsense practices to reduce water demand. In Utah, though, we have made few meaningful attempts thus far to slake our growing thirst. Instead we've focused on increasing the supply of water and just doing things as we've always done them. We certainly haven't looked seriously at reducing the volume of water that we consume because up to now we haven't needed to. An avalanche of empirical evidence suggests that those well-watered days may be coming to a close, however. The evidence—snowpack totals, reservoir capacities, climate models, and population growth predictions—is pretty clear: we're headed for some difficult choices. If Utah continues to focus on supply-side infrastructure projects instead of addressing demand-side reductions in use, for instance, then Utah's fish and wildlife species will be badly impacted. The vast majority of these species depend upon our streams and rivers for at least a portion of their life cycles, and they can't afford to lose another drop of water. The question for Utahns is this: will we have the foresight to act now to face these challenges in a sustainable fashion? Or, without water enlightenment, will we endure the train wreck and then deal with the extraordinarily expensive reactive response to a true water catastrophe?

The two components of the water-population collision are quite well known. Rapid population growth is one variable. A quick look at the U.S. Census reveals some startling statistics about Utah. Many of them require no hyperbole: Utah added more than half a million people between 2000 and 2010, a 23.8 percent population increase that was eclipsed only by Nevada and Arizona. In fact, Utah was the only state to record growth in every one of its twenty-nine counties during the decade just completed.

In the 1990s Utah expanded by a startling 29.6 percent. And over the last half century, in just fifty years, Utah has burgeoned by an extraordinary 210 percent. Only four other states have grown faster over the same period: Nevada, Arizona, Florida, and Alaska. The rest of the booming American West averaged a 156 percent increase in population from 1960 to 2010, well behind Utah's torrid pace. According to the 2010 Census, Utah contained 283 incorporated cities or census designated places. Of those, 223 or nearly 80 percent grew during the last decade. Of cities and places that lost population, only half a dozen are located along the Wasatch Front (the 150-mile urban corridor that stretches along the Wasatch Mountains from Logan in the north to Santaquin in the south).[1] Most of the population losers were small rural towns in the distant corners of the state. When combined with the "Wasatch Back"—that area consisting of the populated mountain valleys east of the crest of the mountains—this greater Wasatch region contains most of the state's population. Except for explosive growth in St. George, most of Utah's 2.8 million residents reside along the Wasatch. And it is here that the greatest demand for water occurs.

Many Utahns attribute this remarkable population expansion to immigration, yet most of Utah's population growth is home-grown: lots of babies. Utah "enjoys" the nation's highest fertility rate, which translates into the country's largest family sizes, greatest number of people per household, youngest population (nearly one-third of Utahns are younger than eighteen), youngest age of first marriage, largest average classroom enrollments, and, not surprisingly, lowest per pupil education spending and highest rate of household bankruptcy in the United States.[2] Immigration is a significant contributor to the state's population to be sure—the majority of immigrants do not arrive from Mexico as is widely believed but from California. In 2011, for example, 85,000 people moved to Utah from other states, 18,237 of them from California alone. The U.S. Census Bureau estimated that an additional 14,465 came from foreign countries.[3] Every one of these new Utahns, whether new by birth or by immigration, needs water. As population grows, water consumption grows with it. By 2060 the state's population is expected to double to more than 5.8 million people.[4] That means that if Utah continues its wasteful water practices, twice as much water will be needed if rates of water consumption remain constant. Water use rates have not remained constant, however. In one study by the Utah Foundation, the state's residents actually increased their daily per capita consumption of water from 285 gallons per day in 1985 to 295 gallons in 2000. Recently that total use has marginally declined, but Utahns are still the second thirstiest people in the nation, trailing only water-drunk Nevada.[5] If Utah's growing population actually increased its per capita water consumption, the head-on collision would be even more severe. We know that the population will continue to grow, but whether the state can provide water without destroying vital habitat for fish and wildlife species remains to be seen.

On the other side of the water-population equation is a declining water supply. Water statistics from a wide range of sources suggest that Utah will face a future with increasingly less water available. The most worrisome of those statistics are those that document a drying climate on the Colorado Plateau and adjacent Great Basin. Computer model after computer model and climatologist after climatologist have noted that Utah has a high probability of drier winters with less snowfall and an earlier melting of that snowfall, which deprives the mountains of the snowpack that keeps the region's reservoirs full long into our typically dry summers. Several recent studies documented a smaller overall snowfall along the Wasatch Mountains, with peak snow depth reached earlier in the spring and melting commencing earlier.[6] What this means is that we have less total snow accumulating and that it begins to melt sooner in the year, long before irrigators and municipalities need the water. A region like the Wasatch Front depends mightily on winter snow reserves for summer water. Even an increase in rainfall wouldn't help much, because the mountain snow saves the water for when we need it most, later in the year.

When the snow does melt and rivers do flow, Utah captures a significant portion of the runoff in a system of reservoirs spread throughout the state and especially in the valleys and canyons along the Wasatch Mountains. While none of these reservoirs is especially large, collectively they store enough runoff for a two-year supply of water for the communities stretching from Logan in the north to Santaquin in the south. Several trends suggest that in the future the two-year storage capacity may become difficult to maintain. One trend is less runoff in general. Diminishing precipitation, smaller snowpack, higher rates of sublimation—snow evaporating directly into the atmosphere—and dry soils that absorb the runoff mean that less water is flowing into our reservoirs.[7] Humans may be partially to blame for these declining stream flows. Dust and soot from our industrial population along the Wasatch is carried by the westerly winds into the mountains, where it lands on the snow. Its dark color absorbs sunlight and hastens the rate of melting and evaporation. Utah's ski resorts understandably worry about declining snowpack. Their solution—making snow—is somewhat controversial and may be aggravating the water supply problem. As more of our mountain ski resorts make snow, they deplete streams and lower aquifer water tables by pumping liquid water and turning it into ice crystals. This snow then suffers the same sublimation fate as naturally occurring snow: it may evaporate into the air before it melts to fill streams, rivers, and reservoirs. Or it may soak into ever-drying soils and infiltrate back into depleted aquifers. Either way, it may be lost for human use later in the year.

Reservoir capacity is also affected by two other trends in Utah's hydrologic system. One is that earlier snowmelt is often intense and short lived. Rather than melting slowly and helping to maintain water levels in lakes for months with a steady inflow, some mountain streams are flooding intensely but only for a short time, after which they drop to a very low level of base flow. Water managers have to save enough room in reservoirs to handle a sudden and severe flow, while protecting downstream areas from flooding. After the flood, though, they are left with low input flows and a water storage system that will only decline for the rest of the year. Water use for cities, industry, and irrigation is not the only reason for falling reservoirs. Evaporation plays a big part in the equation. A drying climate in Utah will only exacerbate the problem of liquid water in our lakes returning to the atmosphere by way of evaporation. Many experts now estimate that even giant reservoirs like Lake Powell and Lake Mead lose as much water through evaporation as passes through their dams and continues downstream. Many of those same experts, using the evidence articulated above, argue that Lake Powell may never fill again.[8]

The second cause of declining reservoir capacity is sedimentation. Most of Utah's human-made lakes are surrounded by poorly cemented sedimentary rocks and weakly consolidated metamorphic and igneous rocks. As rain, snow, and temperature extremes weather and erode these rocks, they break down into ever smaller pieces that flow downhill into rivers and streams. This sediment—gravel, sand, silt, and clay—then finds its way into our lakes and reservoirs. Dams stop the sediment from continuing downstream, so it piles up in the bottom of these water bodies, displacing water and diminishing capacity. In Utah soft rocks combined with a number of human activities that remove their protective mantle of vegetation—overgrazing, off-road vehicle use, human-caused forest fires—mean that a significant amount of sediment ends up in our water storage system. The Utah Division of Water Resources estimates that sedimentation diminishes reservoir capacity by 12,340 acre-feet per year. Removal of that sediment through dredging is costly and leads to other undesirable environmental impacts, such as the issue of where to put the sediment.[9]

An additional worry involves Great Salt Lake, Utah's inland sea. Its large surface area (when it is full, anyway) greatly moderates weather along the Wasatch Front. As large water bodies do everywhere, Great Salt Lake makes nearby temperatures less extreme, raises humidity, and increases precipitation due to the so-called lake effect. As more of the water that would normally drain into Great Salt Lake is siphoned off for human uses, the lake is starved of its natural inflow. The result is a shrinking lake that becomes ever saltier and has less influence on Wasatch Front weather.

Virtually every inlet stream from the Bear River in the north to the many rivers that fill Utah Lake in the south and then drain into Great Salt Lake via the Jordan River has been diverted to green up lawns, fill ponds, water golf courses, and, of course, provide water for cities, industry, and people. In the process Great Salt Lake becomes less great over time—the lake reached a 50-year low in 2005.[10] The decline is particularly extreme for a shallow lake like Great Salt Lake; a few inches lost in depth translate into miles of surface area disappearing. Without that lake's huge surface area absorbing sunlight, increasing humidity, and reradiating energy, the regional climate is altered. The exposed lake bed is also an additional source of the very dust that blows eastward into the mountains and hastens snowmelt. It also appears that with increased salinity—as the lake level falls through stream diversion and evaporation, it becomes ever saltier—the lake-effect precipitation diminishes. With a smaller Salt Lake, the Wasatch becomes hotter in the summer, colder in the winter, and considerably drier. While largely neglected historically, Great Salt Lake is the subject of intense study and research now. The preliminary consensus is that it is much more important ecologically and hydrologically than previously thought. A significantly reduced Great Salt Lake, experts warn, could have catastrophic effects on northern Utah's weather, long-term climate, and water supply.[11]

Many Utahns, aware of our declining reservoir capacity, propose building more dams and diversions. They argue that we should divert even more water from the streams that drain the Uinta Mountains east into the Green and Colorado River systems and from the Bear River of northern Utah. After flowing through thirsty farmland in Wyoming and Idaho and then back into Utah, the Bear River often limps toward Great Salt Lake in sparse snowpack years. Plans to dam it again and push it into a canal or pipeline that would run the length of the Wasatch Front seem unrealistic at best. If there was enough water to fill the canal in a wet year, the downstream Bear River Migratory Bird Refuge would be starved of its lifeblood: water. Additionally, Great Salt Lake would be denied the water of the diverted Bear River (its largest tributary), further exacerbating the weather and climate changes of a declining lake, as described earlier.[12]

To the east of the Wasatch Mountains a complex system of dams, reservoirs, tunnels, siphons, and aqueducts already diverts an average of 130,000 acre-feet of water per year from the Duschesne and Strawberry Rivers on the southern slopes of the Uinta Mountains, through the watershed divide of the Great Basin and into the Provo and Spanish Fork River drainages, to quench the thirst of the southern Wasatch Front communities.[13] If not diverted, that water would travel into the ever-declining Colorado River system to be fought over by the 35 million users in the likes of Las Vegas, Phoenix, and Los Angeles. Supporters of the diversions quickly cite the legal rights that Utah holds as part of its share of water as guaranteed by the Colorado River Compact of 1922. That seminal and historic water allocation of the Colorado River among seven western states (Utah's share is 11.5 percent) was negotiated when the basin-wide population was a sixth of what it is today and, importantly, when the Colorado River was running unusually high. As it turns out, the Colorado has rarely enjoyed as much water as was allocated in 1922. This was the prime reason the compact was revised in 2007 and again in 2012 with a set of interim guidelines detailing how to allocate Colorado River water in the event of what now seems inevitable: significantly lower river flows and water shortages.[14] This most recent revision involved amending a 1944 treaty with Mexico to deal with diminished reservoirs and river flows.[15] Most long-term climate, watershed, and river flow specialists doubt that the Colorado River Basin will ever be as drenched as it was in 1922, at least not in our lifetimes.

Northern Utah's water development is also guided by the Central Utah Project (conveniently called the CUP) as authorized in 1956 by the Colorado River Storage Project Act. This is the act largely responsible for building many of the huge dams on the Colorado River such as Glen Canyon and Flaming Gorge, which directly affect Utah. Utah water developers and managers continue to cite the CUP as justification for planning ever more water diversions.[16] But even though Utah may have legal access to more Colorado River water, the water may not be there to claim. Ironically, St. George in parched southern Utah is proposing a 150-mile long pipeline from ever-declining Lake Powell that would travel south into Arizona and then back north to the fast-growing Washington County city. St. George champions this wildly expensive pipeline to maintain runaway growth in one of the driest parts of the nation. Recently, economists at the University of Utah deemed St. George's pipeline project absolutely unaffordable. In a letter to the state legislature, the economists estimated that the $1 billion-plus project would generate annual debt payments of $47 million, raise water rates 370 percent, and increase impact fees tenfold.[17] Despite a growing chorus of critics and environmental realities, the Utah Division of Water Resources still promotes the pipeline as sound development and economic planning. Its "Lake Powell Project" website prominently displays a 30-year-old photo of a brimming Lake Powell, full of clear blue water. A far more accurate photo would show the reservoir circled with the unsightly bathtub rings that were exposed when it was only 45 percent full in 2013. This all-time low lake level prompted the Bureau of Reclamation to announce that 2014 would mark the first reduced flows from the reservoir since the early 1960s, when the new dam's gates were originally closed. To think of Lake Powell as a reliable source of water in the future is just irresponsible. Legal and affordable or not, the ethics of these unsustainable water diversions should make anyone pause.[18]

Another water hope is tapping aquifers. Aquifers are permeable beds of rock or alluvium that store water in their pore spaces, like giant underground sponges. By drilling a well into the aquifer, water can be pumped to the surface and added to our water supply. Thousands of such wells have been drilled into Utah's aquifers to augment the state's aboveground water system. Many of Utah's aquifers are directly tied to surface water supplies; as lakes, streams, and rivers decline, their recharge of subsurface aquifers also diminishes and water tables—the top of the saturated zone in an aquifer—also decline. And when water tables fall, wells go dry. Many of our aquifers have very slow recharge rates. We are pumping them much faster than they can be refilled, much like a bank account whose withdrawals exceed its deposits. The United States Geological Survey (USGS) estimates that Utah's aquifers have some 200 to 2,000 times more water stored than is added through annual recharge. This means that with current or expanded pumping rates, we'll exhaust this supply of water rather quickly too. Some call these underground reservoirs "fossil water" because it required many millennia for the water to accumulate. According to the USGS, we're essentially using centuries of water accumulation every year. This too is obviously unsustainable.[19]

Barring another dubious dam-building binge to capture the diminishing runoff from a declining snowpack, the only sensible solution to Utah's water woes is to turn the focus from increasing supply to decreasing the demand for water. It is here that we have the very best chance of avoiding the head-on collision between water availability and population growth. Hitherto very little has been done in the way of promoting water conservation. In fact, we've encouraged its waste. Perhaps in a harkening back to our pioneer roots, Utah has insisted on re-creating Ohio in the desert with green grassy lawns, broad leaf trees, golf courses, pools, and ponds. Everywhere (including freeway medians, commercial street rights of way, parking strips, front yards, back yards, and even vacant lots) we have planted grass and shrubs and trees. None of these plants is native to our region, but we insist on having and watering them. In fact, in some of our cities, property owners who don't landscape in thirsty greenery face citations and fines. Not only do our yards and businesses and public spaces have to be planted and heavily watered, but we are encouraged to pour petrochemical fertilizers and herbicides on them so that they are unnaturally and uniformly green and homogenous. Overuse of those same chemicals results in them being carried into streams and lakes, where they compromise water quality. And in the drier future water quality may become as much of a concern as water quantity, because reduced stream flows mean that pollutants gather in higher concentrations.

We encourage this massive water waste not only by mandating exotic green landscaping but by making the water needed to maintain it nearly free. Throughout much of the Wasatch corridor the primary culinary system of potable water is paralleled by a secondary water system that is unmetered and flows in seeming abundance. For a remarkably cheap flat rate—usually 0.1 percent of their property value, as is the case in Ogden, for instance—residents can use as much secondary water as often as they want.[20] And use it they do. They pour it on their lawns and landscaping daily for hours at a time. They fill up artificial ponds and creeks and shoot it over waterfalls and up through fountains. They wash their cars with it and let it run in torrents down the roadside gutters. It doesn't matter, though, because the price never changes no matter how much is used. You can grow algae and raise frogs in your yard, and the flat rate for secondary water will always be the same. This unmetered and unchecked water encourages prodigious waste.[21] Some water watchers have noticed that water providers actually promote water waste; it keeps them employed, financed, and building new projects. If you used less water, they would make less money; and if demand fell low enough, they'd be out of a job.[22]

The fees for culinary water only marginally encourage less waste. Most households pay a flat rate for the first few thousand gallons that they use and then a slightly higher rate for the water that they use beyond that first allotment. On its website the Utah Division of Water Resources prominently lauds the fact that Utah has nearly the cheapest water in the United States. Despite being in the second driest state in the nation in terms of overall annual precipitation, our water is significantly cheaper than in states where precipitation falls far more often and more abundantly. Even in Minnesota, land of 10,000 lakes, residents pay more for their culinary water. Part of the reason for the low cost is that Utah's water is reasonably clean and doesn't require significant treatment to meet water quality standards. Additionally, many water utilities in the state—usually termed water conservancy districts—avoid directly charging for water consumption by instead collecting revenue from property taxes.[23] These policies keep water rates unrealistically cheap. The low water cost is also in large part due to federal taxpayer supported reclamation projects that deliver water to Utahns even though we didn't have to pay for their construction. Many of the dams, canals, and aqueducts that provide the state with irrigation and drinking water were engineered, built, and, importantly, financed by the residents of the entire United States. If Utahns actually had to fund these massive projects and their maintenance, we'd likely be far more careful with our water and price it accordingly.

Hitherto Utah hasn't taken water conservation very seriously. Even while noting that Utah is the second driest state in the union and that Utahns have the second highest water use rate in the nation, the Utah Division of Water Resources proposes only modest conservation measures.[24] The State of Utah's official water conservation goal is to reduce the year 2000 water use of 295 gallons per capita per day (gpcd) by 25 percent by the year 2025, without any other water conservation savings thereafter. This 1 percent reduction of water use per year is considerably less than surrounding states have managed to accomplish. And with a population expected to double by 2050, a 25 percent reduction of water is only half the rate of population expansion. Put another way, we will need half again as much water as we're currently consuming just to meet the proposed future water demand, even with reduced usage. The base water use for Utah's water conservation goal was 295 gpcd from the year 2000, significantly higher than most parts of the country and nearly twice the national average of 155 gpcd, according to the USGS. Despite such high use, the state's water consumption goal for 2025 is a very modest reduction to 221 gpcd, which is still well above the United States average.[25] Unless additional sources and storage areas are found or constructed, Utah will need to cut its water habit by a lot more than 74 gallons per person to avoid severe water challenges by mid-century. It's either that or a far more unlikely scenario: slow population growth.

Given the enormity of the potential water-use crisis, what can be done? A better question is: what must be done? In the water supply-demand equation, virtually nothing can be done about supply. Utahns are unable to make it rain and snow more. We cannot increase snowpack or decrease melting. We've built all the easy reservoirs and diversions. To build more would be recklessly expensive and negatively impact the environment. So that leaves modifying demand. And a lot can be done in this area. Some of the conservation measures that reduce demand are easy to accomplish. For culinary water, low-flow shower heads, toilets, and faucets are an inexpensive start. Water-conserving clothes washers and dishwashers are more costly but represent an easily implemented second effort.[26] More expensive but involving tremendous water savings benefits is rehabilitation of the potable water infrastructure along the Wasatch Front. In many of our communities the tanks and pipes and valves and meters that deliver water are decades old. In some neighborhoods they date back to the nineteenth century. If Utah's antiquated plumbing is anything like some of the oldest systems in the United States, it is estimated that the loss of water through leakage may actually be greater than the amount of water successfully delivered. Updating and expanding this vast waterworks won't be cheap, though. In a 2012 study, the various public agencies charged with ensuring water quantity and quality in Utah estimated the cost of upgrading the region's water infrastructure at a staggering $13.7 billion. As water availability declines and its price inevitably increases, however, that may be a necessary investment.[27]

An additional remedy to discourage high water demand involves rate incentives. Water use and especially water waste should be expensive. Progressive and true-cost water pricing are essential. Not only should water users have to pay more for using more water, but the price must progressively increase to discourage high use. Twenty thousand gallons of water used shouldn't simply cost twice as much as ten thousand gallons used. It should cost a lot more. In parts of Utah, though, the more water someone uses, the cheaper it gets.[28] A second part of the fair water pricing scheme must include the true cost of delivering safe, reliable drinking water to users. It must reflect the actual not the subsidized cost of the water infrastructure, its maintenance, and investments for future water system improvements. All along the Wasatch Front municipalities have coasted on the water investments made by previous generations, thereby keeping water rates artificially low for decades. Water users have taken these low prices for granted. When system-wide rehabilitations and upgrades are necessary, they balk at increasing rates to pay for them. Low rates, like so many of Utah's water expectations, are unsustainable.[29]

The greatest use of Utah water, however, occurs not in the tubs, toilets, and sinks of our houses but outside in our yards and farms. Fully 90 percent (80–85 percent for farms and 6–8 percent for residential landscaping) of the state's limited water is spent to grow plants, most of which would never be found in an arid and semiarid region such as ours. The misuse of this water deprives Utah's native plants and animals of the water they require for survival and has long-term negative impacts on environmental health. Here too much can be done.

A better match between Utah's climate and its residential landscaping doesn't have to mean gravel yards and sagebrush shrubs, as many fear. An entire category of drought-tolerant plants—xerophytic plants to botanists—can be planted to save water and bring appreciable beauty to our yards and landscaped areas. Many desert southwest cities such as Las Vegas, Phoenix, and Tucson have enjoyed significant water conservation success and great-looking landscaping by choosing climate-appropriate plants.[30] In town a number of strategies can significantly cut outside water consumption. Many households have successfully employed rainwater capture and cisterns for yard watering. Moisture-sensing sprinkler systems now "know" when and how much to water. And "smartscaping" uses fewer plants for maximum impact.

Agriculturally, the state's farmers may have to choose less-thirsty crops when faced with a declining and more expensive irrigation supply. Water-loving alfalfa, for example, should be grown somewhere else. In the fields drip irrigation—the system of tiny pipes that deliver water directly to a plant's roots—should replace wasteful flood irrigation. With a drip system, soil moisture sensors can detect the exact moment that plants need water and then deliver small amounts, quickly utilized by the plants before evaporation sends it into the atmosphere. Fertilizer is often added to the water. These "fertigation" systems prove to be very precise and plant friendly.

On a larger urban scale, cities may divert storm water to catchment basins and then later utilize it for parks and sports fields. Additionally, reclaimed sewage effluent may be used for irrigation and decorative water features. When planning for future development, urban water strategists have saved copious quantities of water by employing higher-density urban development with smaller lots, less overall landscaping, and community green spaces and gardens. The benefits of this "new urbanism" stretch far beyond water savings as well. Urban sprawl, air pollution, cheaper utilities, and increased sense of community are just some of the advantages enjoyed through the techniques of new urbanism. Finally, in Utah the unmetered secondary water system must be changed from a one-flat-rate-buys-all-the-water-you-can-waste to a system that rewards conservation and discourages overconsumption. This conversion will accrue costs up front for sediment settling ponds, debris filters, and water meters but will save much water and money in the high-demand, low-supply future for the state's water.

In a larger perspective, our entire water paradigm must be changed in Utah. And the best way to do this is through a sustained and serious water education program. We must become more water wise right now or else face a very precarious future. From elementary school classroom to legislative floor to bully pulpit, Utah's leaders need to preach the gospel of water conservation. For it is only when the state's citizens finally realize the magnitude of the water challenge that we may face, and the pragmatic ways to avoid it, that we can move toward a consensus in water conservation. Education alone likely won't be enough. Water law must be reformed. The principle of "a priori water rights"—first in time, first in right—that allows wasteful and irresponsible water practices only because they are senior rights must be amended. State laws that allow water rights holders not to use their water—also known as a "non-use water permit"—must be extinguished when the water rights holders are not putting their water to use, as is the case with some canal companies that have a surplus of unused water. Remarkably, in Utah there are no instream rights for fish, plants, and wildlife. Other organisms must have equal legal footing when it comes to the allocation of a commodity as precious as water.[31] Of course, nothing is more challenging than reallocating water in a desert. And we will most definitely have a water crisis if a large volume of our water is redistributed to nonhuman uses—instream rights, aquifer recharge, wetland restoration, and letting the rivers flow again into Great Salt Lake—and we do nothing to reduce our growing demand. This really is the key: reducing consumption and waste of water.

Finally, water needs to be priced like the invaluable resource that it is. Like so many of the habitual and hopeful expectations of water use and availability, this will be hard to change. But not to make the adjustments that guarantee a safe and adequate supply of water to all uses and users in Utah is more than irresponsible; it's unsustainable and unethical. Everyone and everything that requires and enjoys water deserves so much better. We may enjoy a wet year and full reservoirs every now and then, but eventually population growth and our water use practices will cross paths in such a way that there isn't enough to go around if we don't make changes soon. Utah would be wise to plan for this future scenario now, before broader environmental problems envelop us. And if we take these steps, our children's children can enjoy a water legacy that guarantees high-quality water for all. They have that right, and we have that obligation.

Notes

1. United States Census Bureau, "State and County Quick Facts," 2010, http://quickfacts.census.gov/qfd/states/49000.html.
2. United States Census Bureau, "Selected Social Characteristics in the United States," 2007–11, http://factfinder2.census.gov/faces/tableservices/jsf/pages/productview.xhtml?pid=ACS_11_5YR_DP02&prodType=table.
3. United States Census Bureau, "Geographic Mobility/Migration," 2011, http://www.census.gov/hhes/migration/data/acs/state-to-state.html.
4. Governor's Office of Planning and Budget, "Demographic and Economic Projections," 2012, http://governor.utah.gov/dea/projections.html.
5. Utah Foundation, "Utah Water Use and Quality," August 1, 2004, http://www.utahfoundation.org/reports/?page_id=331.
6. D. P. Bedford and A. Douglass, "Changing Properties of Snowpack in the Great Salt Lake Basin, Western United States, from a 26-year SNOTEL Record."
7. Utah Rivers Council, "Crossroads Utah: Utah's Climate Future," First Edition, 2012, http://www.utahrivers.org/wp-content/uploads/2012/10/Crossroads.pdf.
8. Bureau of Land Management, "Climate on the Colorado Plateau," May 11, 2011, http://www.blm.gov/ut/st/en/prog/more/CPNPP/Historic_Climate_Conditions.html. See also chapter 10 by Annette McGivney herein.
9. Utah Division of Water Resources, "Managing Sediment in Utah's Reservoirs," March 2010, http://www.water.utah.gov/Planning/ReservoirSediment/Managing%20Sediment%20In%20Utah%27s%20Reservoirs1.pdf.
10. United States Geological Survey, "Great Salt Lake—Lake Elevations and Elevation Changes," January 10, 2013, http://ut.water.usgs.gov/greatsaltlake/elevations/.
11. D. P. Bedford, "The Great Salt Lake: America's Aral Sea?" 8–19; D. P. Bedford, "Utah's Great Salt Lake: A Complex Environmental-Societal System."
12. Utah Division of Water Resources, "Bear River Development," August 2000, http://www.water.utah.gov/brochures/brdev.pdf.
13. Provo River Water Users Association, "Provo River Project Features," 2000–2013, http://www.prwua.org/provo-river-project-features/.
14. U.S. Department of the Interior, Bureau of Reclamation, "Colorado River Interim Guidelines," November 2007, http://www.usbr.gov/lc/region/programs/strategies/FEIS/index.html.
15. U.S. Department of the Interior, "Minute 319," November 20, 2012, http://www.doi.gov/news/pressreleases/secretary-salazar-joins-us-and-mexico-delegations-for-historic-colorado-river-water-agreement-ceremony.cfm.
16. Central Utah Project Completion Act Office, "The Central Utah Project—An Overview," 1992–2012, http://www.cupcao.gov/TheCUP/overview.html.

17. KSL.com Utah, "Lake Powell Pipeline Could Quadruple Water Costs, U. Economists Say," October 16, 2012, http://www.ksl.com/?nid=148&sid=22582832.
18. Utah Division of Water Resources, "Lake Powell Pipeline," 2006–12, http://www.water.utah.gov/lakepowellpipeline/projectupdates/default.asp.
19. USGS, "Ground Water Atlas of the U.S.," February 2009, http://pubs.usgs.gov/ha/ha730/ch_c/C-text3.html.
20. Most of these recipients and secondary water managers have no idea how much water they are using because the water is unmetered. Though the Weber Basin Water District just installed 1,500 meters in its secondary system, this is a small fraction of its total delivery system for secondary water. The meters did show a 100 percent overuse of what was expected.
21. Utah Division of Water Resources, "Municipal and Industrial Water Use in Utah: Why Do We Use So Much Water, When We Live in a Desert?" December 29, 2010, http://www.water.utah.gov/Reports/MUNICIPAL%20AND%20INDUSTRIAL%20WATER%20USE%20in%20UTAH.pdf.
22. Dan Schroeder, "Why the Government Wants You to Waste Water," Lecture, Basin and Range: Water Talks, January 17, 2013.
23. Utah Division of Water Resources, "The Cost of Water in Utah: Why Are Our Costs So Low?" October 27, 2010, http://www.water.utah.gov/Reports/The%20Cost%20of%20Water%20in%20Utah.pdf.
24. Utah Division of Water Resources, "Long-term Water Supply Outlook," April 2012, http://www.water.utah.gov/waterconditions/WaterSupplyOutlook/default.asp.
25. *Salt Lake Tribune*, "Conservation Report Card: Utah Trying to Cut Use, But Still a Top Water Guzzler," November 9, 2009, http://www.sltrib.com/news/ci_13750549.
26. Utah Division of Water Resources, "2009 Residential Water Use," November 3, 2010, http://www.water.utah.gov/Reports/RWU_Study.pdf.
27. *Deseret News*, "Utah's Thirst for Water Comes with $13.7 Billion Price Tag," October 26, 2012, http://www.deseretnews.com/article/865565407/Utahs-thirst-for-water-comes-with-137-billion-price-tag.html.
28. Western Resources Advocates and Utah Rivers Council, "Water Rate Structures in Utah," January 2005, http://www.westernresourceadvocates.org/media/pdf/Utah%20Water%20Rate%20Analysis%20-%20300dpi.pdf.
29. Utah Rivers Council, "Free Market Water," 2012, http://www.utahrivers.org/collecting-taxes-to-encourage-water-waste/.
30. Utah Division of Water Resources, "Municipal and Industrial Water Use in Utah," December 29, 2010, http://www.water.utah.gov/Reports/MUNICIPAL%20AND%20INDUSTRIAL%20WATER%20USE%20in%20UTAH.pdf.
31. For a detailed discussion of the issues connected with water rights, see chapter 13 by Daniel McCool herein.

2

The Miracle at the End of the Line

STEPHEN TRIMBLE

WE HAVE A HOUSE in southern Utah at the foot of the high plateaus. At an altitude of 7,000 feet, on a red rock mesa at the brink of wilderness—we live in an unlikely place to flip on the lights, take a hot shower, flush a toilet, bake a pie. The village of Torrey scatters across a rocky bench a mile away, and we sit at the end of the line for the town water system—the last house reached by the system of pipes that starts some 17 miles and 3,500 feet above us on Thousand Lake Mountain.

I turn on the tap. Water pours out—so cold in winter that it hurts your teeth, not much more than one degree away from ice chips. In summer the water seems miraculous, life-sustaining, restorative—and tastes like a mountain stream swirling between rocks, past nodding columbines, and right into our sink.

In southern Utah water arranges the location of towns. A cluster of communities in western Wayne County survives where creeks and springs from Boulder Mountain and Thousand Lake Mountain meet the Fremont River.

Downstream from this passage between the mountains, the river drops eastward into the desert, through the slickrock wildlands of the Waterpocket Fold in Capitol Reef National Park, through the Mancos Shale badlands around Factory Butte, and finally to the junction with Muddy Creek at Hanksville, where the Fremont becomes the Dirty Devil River.

As in most political provinces, nonsensical straight lines outline three sides of Wayne County. But if we adopt the perspective of John Wesley

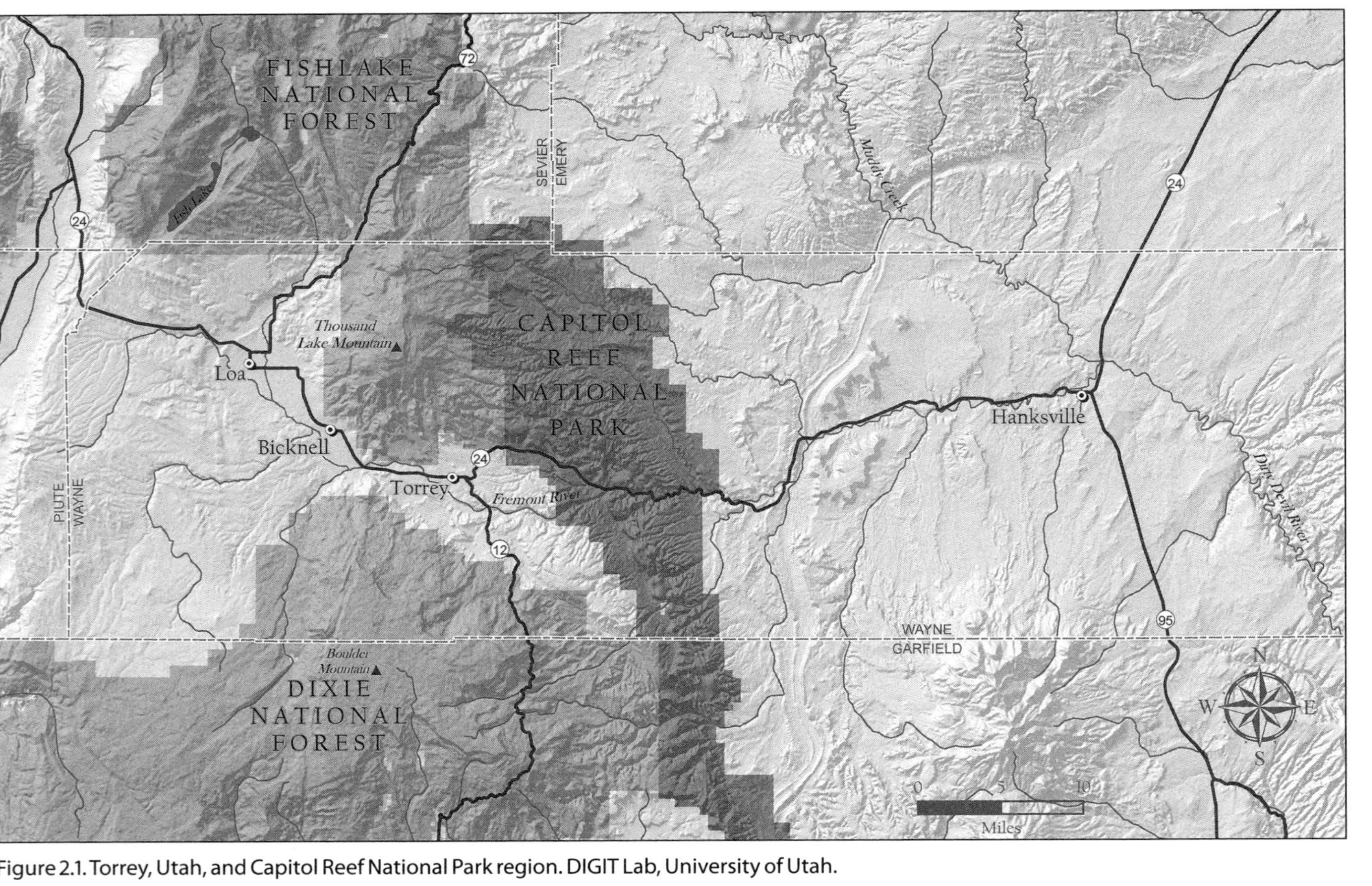

Figure 2.1. Torrey, Utah, and Capitol Reef National Park region. DIGIT Lab, University of Utah.

Powell—who proposed managing the arid West watershed by watershed—the Fremont and Dirty Devil watersheds match the core of Wayne County's straight lines reasonably well. The source of the Fremont in Fish Lake lies just to the northwest, and the lower Dirty Devil reaching to its confluence with the Colorado lies just to the southeast.

Beyond the mountains, for nearly fifty miles downriver east of our town of Torrey, only three villages exist. Eleven miles downstream the historic community of Fruita has morphed into Capitol Reef National Park headquarters, where a few families of park rangers live. Another twenty miles downstream lies Caineville, population 27 (and that's only "if everybody is in town," as Caineville farmer Randy Ramsley says).[1] At the junction of the Dirty Devil and the Muddy, Hanksville, population 220, is the last outpost of village life in Wayne County.

The county runs to the east an additional forty miles, ending at the Green and Colorado Rivers. A single family lives out there, at Robbers Roost Ranch, the only inhabited spot in the sixty-five miles between Hanksville and Moab as the raven flies.

This is empty country both because it's dry and because it's a maze of slickrock mesas, slot canyons, moonscape badlands, sand deserts, and remote plateaus. Average annual rainfall at Hanksville: 5.5 inches; at Torrey, 7.5 inches.

Back between the mountains, in the Torrey Valley, dozens of new homes nestle into the mesas outside of the villages. Between Boulder Mountain and the Teasdale Fault, homeowners have good luck drilling wells into the Navajo Sandstone. North of the fault they hit water that's not so sweet or find no water at all. They need city water; if they can't get it, they must resort to hauling water. We're lucky. We turn on our tap, and Torrey Town Water pours forth.

Water surfaces repeatedly in any discussion of sustainable futures for Utah communities, for Torrey and for all of rural Utah, with the inexorable energy of an August flash flood.

How much growth will come to Torrey? How much growth do we want? How much growth can we sustain?

The answers to these questions aren't unique to Torrey. The story of my village symbolizes the story of small towns in most of Utah—indeed, small towns across the arid West.

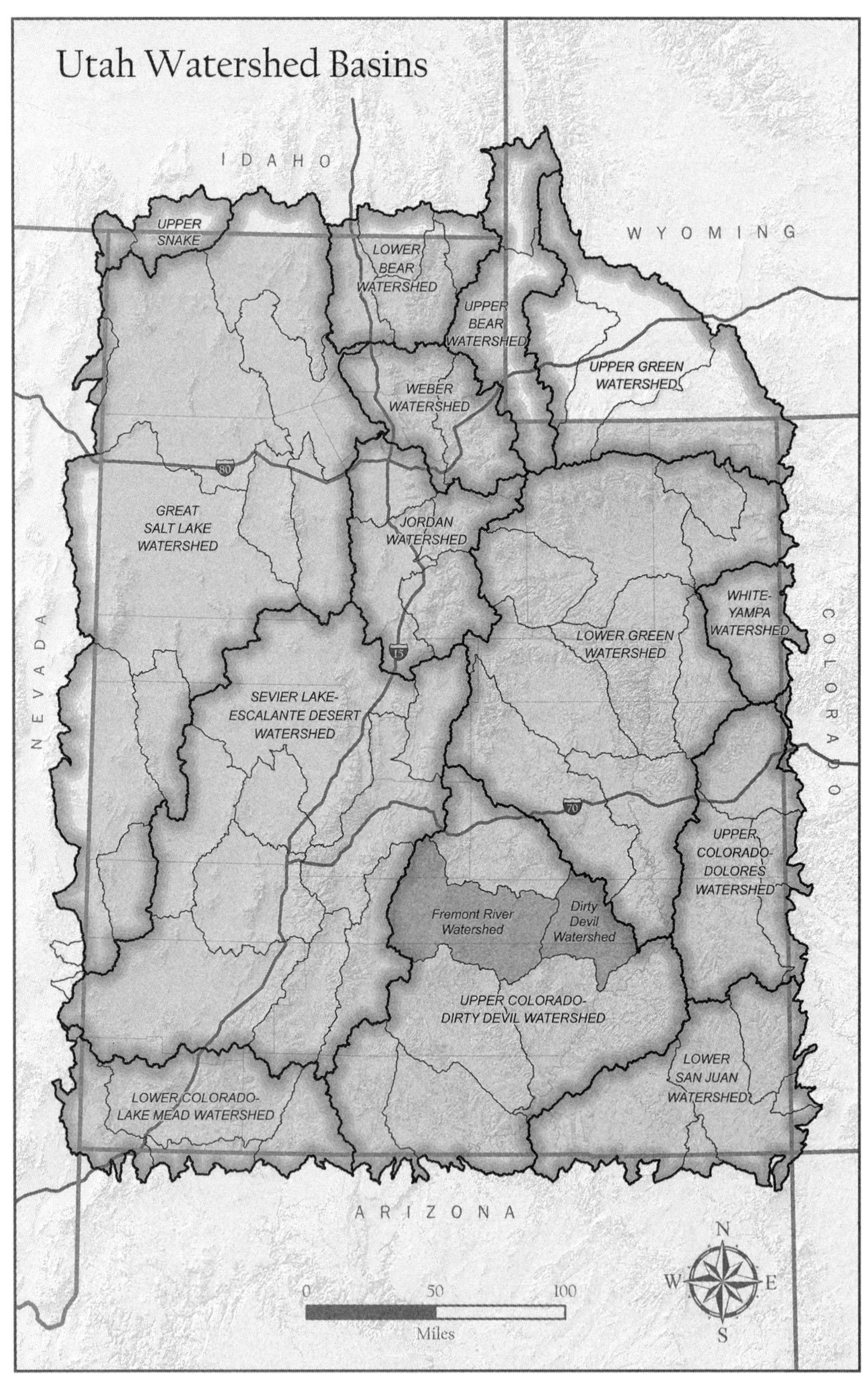

Figure 2.2. Utah watershed basins. DIGIT Lab, University of Utah.

Still, our Thousand Lake Mountain water is special—most of the time untreated, straight out of springs above 10,000 feet. How has our community done this? What does it take for my tap to flow?

I wondered how this happens so dependably. And so I asked the old-timers in Torrey for history and Utah water experts for data. What's the story of our water? What does it take to have that miracle of mountain water tumble into our glasses?

The answers to my simple questions remind me how contentious water remains in the arid West—and how complex our relationship with this fundamental resource. As so many people said to me about Torrey's water supply: "Water is touchy. Water is complicated." "We have one of the most complex systems of any small town in southern Utah." "It's a marvel."

In these conversations, we quickly turn emotional, personal, unforgiving—and generous. I hear phrases like "crooked as a snake in the grass," "stupid," "out of touch," and "liar." I also hear: "He has the biggest heart." "He's such a hard worker." "He knows the system better than anybody."

The old saw "Whiskey is for drinkin' and water is for fightin'" remains more true than we might wish. Water isn't just life in the desert; water is the x-axis for a graph of multigenerational grudges.

Mormon pioneers arrived in Wayne County in the 1870s. They came into the upper Fremont Valley—Rabbit Valley—to graze cattle in the most extensive meadows found in the 2,500-square-mile county. Once they built ditches and canals, they could irrigate crops; reservoirs on the Fremont River followed, to store water in dry seasons and control the river and its tributaries in summer flood season.

The settlers established their first towns. As families grew and new people came into the community, they scouted downriver, pushing out from Fremont, Loa, Lyman, and Bicknell.

Pioneers located Torrey where Sand Creek flows down from Thousand Lake Mountain and out into the Fremont Valley. The Fremont could provide irrigation water—but first a canal had to be built to bring the Fremont up onto the bench where folks had built their town. Sand Creek—cleaner and clearer because it ran right off the mountain—could provide drinking water.

During the 1880s families trickled in to live on the flat bench fifty feet above the river. But their plans didn't go smoothly. As Sarah Smith tells us in the classic 1953 county history *Rainbow Views*, settlement in Torrey had reached a threshold by the end of the decade—and the church organized a new branch to serve the new town. "These people expected water to be brought from the Fremont River by a canal," Smith wrote. The canal was surveyed and ready to go. "But when nothing was done about digging it, most of the people moved away, and the branch was discontinued."[2]

No water, no Torrey. The history of Torrey is the history of the town's water. In the 1890s water rights to Sand Creek changed hands several times, and people began to move back to the Torrey bench. As Sarah Smith says, "The idea of bringing water from the river was always in their minds. More water meant more farms, gardens, and homes."

Over time, those who tired of relentless hardship and flooding downstream returned up-county as far as Torrey to live. Finally, the Torrey Irrigation Company incorporated in 1902; it took more than twenty-five years to improve the canal, ditches, and pipeline system incrementally to bring irrigation water to the villagers and their fields. People worked with teams of horses and got paid with water for their work. Scottish stonemason Robert Peden labored for ten years to cut a channel through bedrock that came to be known as the Peden Cut. Improvements to the Torrey ditch followed, generation to generation, including major work in 1977 (new gates, siphon, gauges, and pipe).

For drinking water Torrey residents relied on Sand Creek. Early settlers hauled water in barrels, especially in winter when the creek froze. They tried covered cement cisterns, big enough to last the winter. Not until 1936 did culinary water reach Torrey in pipes, and Anne Snow wrote in *Rainbow Views* in 1953 that "most of the residents now have it in their homes."[3]

Federal grant money paid for that first culinary system—piping water into town from a small reservoir/settling pond on Torrey's outskirts. Thirty years later federal funds allowed the town to upgrade further, shifting from the open reservoir to a 60,000-gallon concrete water tank. By 1968 Torrey's culinary water ran in pipes all the way from Birch and Indian Springs on Thousand Lake Mountain. The Sand Creek Irrigation Company board tired of seeing the creek go dry on the hottest summer days and brought water down Thousand Lake Mountain in pipes as well.

Torrey kept its water system functional for a long time with classic small-town Utah self-reliance. Torrey's Mayor Walt Smith snagged the rights to Chaffin Spring, purchased for a song in the 1950s—and today this huge spring bubbling right out of the cliff is the town's best producer. Mayor Jay Chesnut finally brought the entire water system into pipes in the 1960s. Both officials took a drubbing locally for spending any money at all.

Jay Chesnut's son, Scott, "grew up with the water system. I worked on it forever. Before they got all these pressure release valves in there, it would airlock coming off the mountain and you couldn't get any water. You had to tend it nicely or it wouldn't produce. It's temperamental."

Working on those lines took Torrey crews up to Thousand Lake Mountain, horsepacking the pipe to the four remote springs that serve the town: Birch, Indian, Sulphur, and Chaffin. At first they ran pipe across the ground surface, right over the talus, until they could get Forest Service easements that allowed them to bury the line. When they succeeded, Torrey's culinary water finally ran in pipes from mountain spring to village home. A proud local catchphrase exulted in their success: "Our water never sees the light of day."

After the pipes went in, bypassing the settling pond, folks also said they had to stir some red dirt into their water to give it the familiar density that tasted right.

In the early years rural Utahns organized their water systems cooperatively, often under the direction of the local bishop of the LDS Church. Together they decided who had the rights to shares in a given water system. As communities grew and demand increased, water companies took over from the church hierarchy; municipalities claimed their own rights. In a small town these overlapping constituencies drew their leaders from the same tiny pool of citizens, so a single person could be mayor, bishop, and president of the local irrigation company in turn—and repeatedly.

With growth, limited water supplies like the Fremont River had more claimants than water—and no rules as yet. Squabbles ensued.

In 1935 the Hanksville Irrigation Company filed suit against upstream Fremont River water users.[4] Judge Nephi Bates of Utah's Sixth District Court wrote a decree that gave Hanksville and the other downstream users priority rights. And so Torrey irrigators could divert water only if the people downstream in Fruita, Caineville, and Hanksville had taken their decreed right.

Floods and drought—the imperious unpredictability of desert streams—confronted the new water commissioner hired to administer the Bates decree, Freeman Tanner. How could Tanner monitor fair appropriation of water when the measuring weirs kept washing out? How could he be fair in midsummer, when the river simply couldn't deliver enough water to meet everyone's needs?

Wrangling continued for years. Capitol Reef superintendent Charlie Kelly reported on one 1950 meeting in Fruita to try to work out conflicts over shares that "ended in a brawl."[5] He walked out. In 1963 the Torrey Irrigation Company challenged the Hanksville company's actions in court.

Wayne County residents have longed dreamed of a dam on the Fremont near Torrey to store water and control floods. This proposal resurfaced energetically in the late 1980s. Both the National Park Service and Bureau of Land Management opposed the idea because of potentially disastrous effects on riparian habitat and water quality. Local proponents simply submitted the application without the critical agency comments. Federal regulators rejected the application when they spotted these shenanigans. The dam project was killed for good in 1991.

Back in Torrey town, water and development generated new showers of sparks during the first decade of the twentieth century.

Fred Hansen was serving as Torrey mayor at the end of a five-year drought in 2004. The town decreed that no new home could connect to the city water system unless the owner had an existing meter. At the time Hansen told me, "I'm sorry this drought caught up with us while I'm mayor, but it's good for the community. It woke us up. We just quit selling water; we can't sell something we don't have."

This became one of those times when community goodwill broke down. The Torrey water system was leaky and deficient, true. But Hansen also had a conflict of interest: he owned a significant number of Sand Creek Irrigation shares and wanted to make sure that overflow from the town water system kept running into Sand Creek Irrigation Company pipes. And so he refused to sell water to property owners who had been promised meters and hookups—and water. Irate, the landowners hired lawyers. Four painful years led to a new town ordinance in 2008 that allowed people who wanted to build on their land once again to purchase a water hookup in Torrey.

We have reveled in wet years, suffered through dry years, and lived complacently through "normal" years. Some Torrey citizens agree with Fred Hansen that water will limit our growth. Others object, saying that the springs produce plenty of water; if we need more, we can get more with increased delivery rates and more care taken to improve efficiency.

To explore these contrasting visions, I looked for a wider context, for commentary from our statewide experts on the interplay between our increasingly congested future and our middle-of-nowhere agricultural heritage.

"The bottom line for rural America is that affordable drink of water. It needs to be reliable, it needs to be safe. It needs to be clean." Dale Pierson, executive director of the Utah Rural Water Association, starts his stock sermon with these basic principles. When I visited with him in his office in Alpine, he also talked me through the big issue for small water systems: regulations in place since passage of the 1977 Clean Water Act and 1974 Safe Drinking Water Act. Torrey has to answer to the Environmental Protection Agency, to the Drinking Water Board for culinary health, to the Water Quality Board for septic system and wastewater safety. This is a daunting task for a village with a limited budget for replacing crumbling infrastructure.

Pierson believes in those regulations. As he says, "Water is life. Air is only a little more important. Our physical health, our environmental health, and our economic health all come down to having good water systems." The regulations ensure that health. But the rules work better in cities than in communities with only a few hundred water meters; the smallest towns simply cannot afford to upgrade to a system that provides not only tap water but adequate flows for fire hydrants.

Dale Pierson helps small communities find money for modernizing their water infrastructure. Without those state and federal grants, he notes, "throughout rural America, if there aren't ways of getting funding you'll drive economic development to the big cities, and rural America will stagnate."[6]

Not just stagnate but wither. Cathy Bagley—intimately familiar with the changing demographics of the community because she has made her living as Torrey's primary real estate agent since she moved here in 1970—notes: "It's always been a real sign of success if you could stay here. It didn't matter if you worked in a gas station. That's what success meant." Wayne County families need water systems to create that chance for success.

Torrey has a population of 180. A few hundred additional people scatter across the mesas and flats just out of town, in small subdivisions and standalone homes—hooked up to town water. Water is limited but may not be the primary limiting factor for growth. Cathy Bagley says:

> The wind blows, but we have good water—plenty of water.
>
> What is the limiting factor for Torrey? There's just not that many people who want to be here. You drive in—and it's a cute little town, but "Oooh, it's cold in the winter. Ohh, it's windy." It's a long ways from an airport. You gotta bring your money with you; you aren't going to make your living here.

If you don't have a water right when the last shares are distributed, you won't have a water right. Torrey isn't looking for more water. This isn't St. George. No plans exist for a pipeline from Lake Powell. Growth really is limited.

Steve Clyde, a Salt Lake City water lawyer who serves on Utah's Executive Water Task Force, knows all sides of these challenges:

> Torrey probably has enough water for its existing residents—and maybe can even take a little more growth. But you will hit a point where the only place you can find that water is to reach out and take it from agriculture. And if you are going to do that, you are hurting your local economy, and you are going to retire those lands either to subdivide or strip the water from them.
>
> We could, with some change in the law, and some creative public funding, promote on-farm conservation; water that is being inefficiently used on the farms and lost could be salvaged, and that water could be put to use in municipalities. But how do we do that without destroying our ag heritage—which is still a big part of our economy and culture?[7]

That's a real fear—and as soon as this transfer threatens to become reality, anger and fear and misunderstandings run rampant.

Torrey raises water rates to pay for improvements, and users bristle. Torrey charges more for out-of-town users, which complicates conversations about possible annexation of out-of-town neighborhoods. All this in a town where Mayor Adus Dorsey can sit in the town office and point east and west and say truthfully: "Our town ends three blocks that way and three blocks that way."

Dorsey has lived in Torrey for twenty-five years and has been mayor since 2010. He's rounded up money for the latest round of improvements from Utah's Permanent Community Impact Fund Board (CIB): $300,000 in zero-percent loans—to be repaid by Torrey over many years—and $1.3 million in outright grants. The CIB money comes from mineral lease royalties returned to the state by the federal government—and can only be spent in communities "socially or economically impacted, directly or indirectly, by mineral resource development on federal lands."[8] Luckily for Torrey and its neighbors down the road, that clause probably applies to every small town on the Colorado Plateau.

Now meters measure output on each mountain spring. Additional water tanks, isolation valves, looped pipes (to keep water moving and increase pressure), and a water treatment unit all will be in place by the end of 2013.

Regulations nudge system operators toward chlorination. Dale Pierson, at the Rural Water Association, pushes towns toward water treatment: "It's incumbent on the water operator to constantly be looking ahead, to have a good plan for the future, to look at those needed improvements to meet the regulations. If you run into problems, you have to start treatment."

Torrey's Mayor Adus Dorsey reassures me: "Most of the time our water is untreated. When you say 'treatment' out loud, everybody freaks out, but if we get a big rain or the system gets compromised by a flood, we'll get something into the system and we've got to take it out. Coliform. Woody matter, roots."

In the past Torrey's water operators just "tossed pucks of chlorine in the tanks," to Dorsey's chagrin. The new system will include a treatment facility at Birch Springs to fine-tune the amount of chlorine needed to treat Torrey's drinking water when necessary.

Locals resist. We love our water. We may have the Wayne County wind in full force, but we have good water. Part of Sand Creek Irrigation Company's water right is overflow. Whatever the town doesn't use, the Sand Creek Company gets. And so we don't need to treat all the water, because the overflow goes into the irrigation company pipes. But only those with shares in Sand Creek water can use that system; everybody else must water their lawns and gardens with increasingly valuable Torrey culinary water.

With the new water system, the town will divert water above Birch Springs and treat only the culinary water. Adus Dorsey says: "Once we have this system in place, we'll be able to dial it right in. We'll know exactly how much chlorine we need to put into the system. Now we can do it right—intermittently, scientifically, and safely—instead of guessing."

Those of us who love that springwater hope that he doesn't have to do this too often.

I have my answer. The miracle at the end of the line—the water that fills my glass at our house on the mesa—comes, not surprisingly, with a story in every drop. The tale begins at the top of my local mountain and flows down through everything that makes this landscape my home—geology, culture, economy, climate. Each one of those overarching themes is dynamic—including climate, which may well provide the driving force for the challenging next chapters of living in this increasingly arid land.

My water line runs through more than a century of ambition, anger, courage, and greed. Faith, love, luck, wonder. Pride, pluck, persistence. Cleverness, competition, cooperation—and generations of backbreaking shovel work.

Welcome to Torrey. Welcome to the West. Welcome to the miracle of Thousand Lake Mountain springwater in my glass.

Notes

1. All conversations cited in this chapter took place in 2013 in Torrey, Utah, unless otherwise noted in the text.
2. Anne Snow, comp., *Rainbow Views: A History of Wayne County*, 260.
3. Ibid., 261.
4. Miriam B. Murphy, *A History of Wayne County.*
5. Bradford J. Frye, *From Barrier to Crossroads: An Administrative History of Capitol Reef National Park, Utah.* See especially chapter 13: "Water Rights and Water Quality at Capitol Reef."
6. Ibid.
7. Steven Clyde, water lawyer with Clyde, Snow, and Sessions, personal interview by Stephen Trimble, Salt Lake City, February 12, 2013.
8. http://jobs.utah.gov/housing/cib/cib.html.

3

Bear River

Learning from a River That Closes Our Circle

CRAIG DENTON

To be worthy of recognition and protection, a river needs a statuesque public persona. The Colorado has one. So does the Green. Because they do, Wasatch Front residents revere them—from a distance. The Bear River doesn't get similar respect, although it should, because our local river is a closed system. Beginning in Utah and emptying into Great Salt Lake, it lives in our neighborhood. We "own" it, even if we don't have the rights to all the water that flows in it. In a closed hydrologic system, the mandate to recognize the river and husband it is more pertinent, because our river isn't just passing through like the Colorado and Green. We can't brutalize the Bear and send our offal downstream, out of sight and out of mind. If we do, we imperil our own neighborhood.

Opening the Circle

The Bear is Utah's homegrown river. Moreover, it's a river in its own right, not a tributary. After its headwaters gather snowmelt in glacial basins and lakes in the northwestern Uinta Mountains, the Bear River follows its gravitational lodestar due north. Plunging madly over its first five miles, it begins to meander on the Utah-Wyoming border. It dips briefly into Utah again then back to Wyoming before it slips quietly into Idaho.

The Bear once was a mere tributary of the larger Columbia River system. But about 34,000 years ago at a place now called Soda Springs, Idaho, a volcanic upheaval oozed lava into its path, blocking its flow north, and the river became youthful again, christened with an identity of its own. At 250 miles out with magma-turned-to-basalt blocking its course, the Bear River made a hairpin turn and headed back to its birthplace. With a total length of 500 miles, the Bear River empties into Great Salt Lake, only 75 miles west of its headwaters.

The voyage seems laughably contorted, as if the river couldn't make up its mind. To human beings coming upon its banks, however, the river became a critical navigational path. Native Americans pressed bare feet then sandals and moccasins into its soft dirt beginning 15,000 years ago. Trappers followed those paths in the 1820s and 1830s into the mountains of the Bear River Range, the northern most extension of the Wasatch Mountains that splits the Bear River Basin. The year 1845 brought the first of the explorer-cartographers, John C. Frémont, to the banks of the Bear, where he was one of the first to envision an inland empire.

Loners most of the year, the mountain men followed the meanders of the Bear in search of personal wealth. Emigrants to the Bear River Basin in the 1840s came for another reason: community. The Bear River was a signpost for them. Wagon trains meeting the Bear in Wyoming and Idaho followed the Oregon Trail adjacent to the Bear northward. When other emigrant groups looked westward, entrepreneurs built ferries across the Bear to move that traffic. In the 1860s towns wrestled to become the railroad transportation center of the Intermountain West. Corinne saw its site on the Bear close to the river's delta on Great Salt Lake and near the site where the spike linking the transcontinental railroad would be driven as fortuitous. No doubt good geographic fortune would enable that city to declare itself the "Chicago of the West." It wasn't to be: dreams for a lively steamboat trade on Great Salt Lake died when the Bear's decreasing flow wouldn't provide enough draft for a boat.

For a while, then, the Bear River reverted to its sole historical role nurturing one of the world's most prolific and rich ecosystems, delivering 1.2 million acre-feet of water annually to Great Salt Lake, 60 percent of the inflow.[1] The marshes and mudflats of Great Salt Lake support more than 260 species of birds. The lake's wetlands host approximately 33 species of shorebirds. From 2 to 5 million individual birds rely on the inland lake at some point during the year. It's such a rich habitat that it's part of the Western Hemisphere Shorebird Reserve Network. Moreover, the natural partnership of the river and lake creates a refuge vital to the Pacific flyway.

What happens at the Bear River Refuge doesn't stay at the refuge. It ripples into the natural histories of myriad other refuges in the flyway, as migratory waterfowl rely on the series of waterholes.

Like all rivers the Bear also carries the natural history of its basin in its watery veins. It slides over hard Precambrian rock in its headwaters, picking up more molecular ions than granular sediment. As it heads north, though, the rock around it becomes increasingly younger, just as the river is rechristened at Soda Springs. When it turned south after igneous rock blocked its way, the Bear began flowing into Lake Bonneville, that vast inland sea in the Pleistocene. Lake Bonneville already was a large body of water, but it grew rapidly when the Bear started feeding it, eventually breaking through an ice dam and momentarily flowing into the Columbia River Basin. After that cataclysm, and as the last ice age began to ebb as yearly rainfall decreased, eventually the river and lake reached an equilibrium.

As temperatures rose the lake shrank, exposing the fine silts and clays that the Bear had deposited. Today the Bear River cuts through those sediments over its last 200 miles. While the river is unnervingly clear in its headwaters region and rushes downhill with breathless abandon and velocity, it takes on an opaque, muddy personality in its lower reaches, as it cuts through the silty deposits left over from the Pleistocene. Unfortunately for the Bear, those lower reaches are where most Utahns live. If drivers passing over the Bear on interstate bridges see it at all, the river appears stagnant and unclean. With that perception comes a parallel question: Why take care of a dirty river so tired that it barely moves?

Tying the Bear into Knots

Imagining something fluid like a river being tied into knots defies common sense, but that's what water law does. Because miners in territorial Utah had to divert water from creeks in order to work their claims, they needed predictable sources of water. The idea of rights to water flowing in rivers and streams grew out of that necessity. Emerging water law carved a private property right out of a public resource: water. "First in time, first in use" became the governing dictum. Those with the earliest water rights took their place at the head of the line when it came time to divvy up river water. The earliest water right filed on the Bear belonged to Wyoming's Myers Ranch in 1862. In Utah the earliest water rights on Bear River water date to the 1880s, having been transferred from the Telluride Power Company and its successor, Utah Power and Light, to the Utah-Idaho Sugar Company and finally to the Bear River Canal Company in Box Elder County.[2]

Water law also required that diverting river water must be "beneficial." Leaving water in a natural streambed wasn't beneficial, again defying common sense, because it wasn't being put to *human* use. Once in place water law allowed a river to become monetized, dividing it into smaller and smaller fractions as more people put more claims upon it. Water rights themselves morphed into a Byzantine menu of options—foundational rights, irrigation rights, storage rights, seasonal rights, senior rights, junior rights, flow rights, quantity rights, natural-flow rights, diligence water, generation rights, stock water rights, to name a few—and water law enriched enterprising lawyers. Eventually the Bear, like other rivers, began to lose its character, becoming nothing more than a medium for conveying water rights to someone downstream.

Water law has set the stage for conflict in the Bear River Basin. Rights to water are entwined with the Old West, founded on land and ideologies cultivated by the first settlers. But the Old West is vanishing in the face of the urbanized New West. Those urban residents whose grandparents aren't buried in small-town cemeteries in Utah would like to see the Bear River put to other, nonconsumptive uses: namely, recreation. But it's hard for them to get a seat at the table because, if they have any rights to river water, they are junior to the more senior consumptive water users.

Equally confounding, water and bloodlines have merged. Families and water rights have become one, so suggesting new uses for water threatens the fabric of families who've used that water for generations to eke a living from the ground. Moreover, water law insinuates itself into cherished mythologies of the Old West. Settlers were told that westward migration was our country's manifest destiny and that they should make the desert bloom. Further twisting the Gordian knot that binds up water, pioneers came to believe in ecclesiastic determinism, the notion that God gave them dominion over nature so that they could be fruitful and multiply. These myths blended into a heady, intoxicating concoction that makes it difficult to change course.

There was one more step in the consecration of Bear River water uses. It is one thing to make a claim on water. It is another to have it recognized in court, which means that most water rights need to be adjudicated. When rivers such as the Bear and Colorado pass through several states, adjudication begins with a multistate compact that formalizes the claims through courts. In 1946 Congress granted Utah, Idaho, and Wyoming the right to negotiate and enter into such a compact. In 1958 the three states signed the Bear River Compact, and a commission was set up to carry out its provisions. Article 1 of the compact states that its purpose is "to remove the causes of present and future controversy over the distribution and use of the Bear River; to provide for efficient use of the water for multiple purposes; to permit additional development of the water resources of the Bear River; and to promote interstate comity."[3]

In effect the Bear River Compact sanctified the status quo and reaffirmed the power of the Old West. Irrigation would be the dominant, best use of Bear River water, and generating hydropower would be a close second. Newer users would have to wait in line or buy those senior rights from willing sellers. If they could purchase a right, they'd have to make that use beneficial in order to exercise it. In most instances courts don't see conservation as a beneficial use.

Threats to the Bear River

As indicated elsewhere in this collection, Utah is the second driest state in the nation, yet it has the second highest per-capita water use, roughly 20 percent more than the average throughout the Intermountain West.[4]

Faced with an exploding population, Utah water purveyors long have looked at the Bear River as a source of water for new development. In 1991 the Utah State Legislature decreed that Utah should begin to develop the 275,000 acre-feet allotted to it under the Bear River Compact.[5] The first step would be to divert 50,000 acre-feet for use by the Weber Basin Water Conservancy District and another 50,000 to the Jordan Valley Water Conservancy District (JVWCD) in the rapidly growing west side of Salt Lake County.[6] While Salt Lake City and the east side of Salt Lake Valley get approximately 60 percent of their water from surface streams supplemented by Provo River water via Deer Creek Reservoir, the west side of the valley doesn't have those kinds of sources flowing out of the Oquirrh Mountains. Its water needs to come from elsewhere. With population projections showing the need for an additional 500,000 acre-feet by mid-century,[7] the JVWCD claimed to need every drop that it can find from existing water rights holders, conversion of gray water, or conservation.

To move water south from the Bear River to Salt Lake and Weber Counties, the state began looking for possible dam sites. Engineers determined that a site near Honeyville in Box Elder County would be the best choice. The Amalga-Barrens area in Cache Valley was viewed as the second best. But an effective coalition of Shoshones, county ranchers and farmers, and Bear River Migratory Bird Refuge officials and conservationists led by the Utah Rivers Council thwarted the siting decision. They argued that a dam and reservoir at Honeyville would bury archaeological remains; displace fifty farming families and threaten water, sewer, and irrigation drainage systems; destroy valuable farmland; cause the refuge to lose 20 percent of its water at a critical time in the summer; and leave a shallow, barren mudflat after the reservoir was emptied by September, a habitat only good for cockleburs.[8] After that pressure the legislature passed a law saying that no dam would be constructed at Honeyville—in the foreseeable future—or at Amalga-Barrens, for many of the same reasons.[9] The idea was floated that perhaps Bear River water could be pumped into an enlarged Willard Bay for storage before it made its way south via a pipeline.

Dams aren't the only threat to the Bear River. Because irrigation is the dominant use in the watershed, farm runoff typically discharges fertilizers (phosphorus and nitrogen), herbicides, and pesticides into the river. The steep banks of the Bear through the lower Bear River valleys in Idaho and Utah remain wet during summer because of increased flows for irrigation. That moisture moves laterally into the soft sediments left from Lake Bonneville, and the banks slough off into the river. This increased sedimentation smothers macroinvertebrates, which provide food for fish and some waterfowl, diminishing biodiversity adjacent to the river, in addition to destroying nesting spots for birds. Plants can't grow on the bottom of the river because sunlight can't penetrate the muddy water, creating a problem for diving ducks seeking food. River flows managed for irrigation upset the natural hydrograph, and insects don't prosper when their aquatic habitat can unnaturally fluctuate several feet in the course of a few hours, leaving them vulnerable to the blistering sun.

Trying to maximize every square foot of land for cultivation, farmers often plow to the riverbank, eliminating riparian habitat. When stretches of the Bear are dewatered for irrigation or hydropower generation, as is the case in Idaho's Black Canyon, or when riverine canopies are chopped down to prevent trees from using up water, species lose connectivity and the biota is stripped of genetic diversity.

Government agencies or Non-Government Organizations (NGOs) can be stymied by a lack of data when they try to monitor the Bear and measure degradation, whether due to excessive sedimentation from upstream erosion, off-the-charts nitrogen load from area farms, or concentrations of herbicides or pesticides. Nancy Mesner, associate professor of watershed sciences at Utah State University, finds this lack of data frustrating. Either there's a "cloud of undifferentiated data" or new data can't be used because earlier researchers, looking at different problems, didn't measure baseline data. In effect much current research has to start from scratch.[10] If those threats aren't enough of a challenge to the Bear River and its stakeholders, they also face the question of climate change and how severe the changes will be. People agree that the southwestern United States will encounter more frequent drought and less frequent but more severe storms because climate in inland locations lacks the mediating effects of the ocean. Utah State climatologist Robert Gillies notes that the methodology employed in regions with stable variables, such as the Pacific Northwest, don't work well for Utah. "We are in a weird situation," Gillies says, because the jet stream drives northern Utah climate while the tropical Hadley Cell influences southern Utah's climate.[11] Nevertheless, "observation records show the snow cover across Utah has decreased 35 percent over the past 50 years, alongside a declining snow depth…this may force us to reexamine some water policies," notes Gillies.[12]

In 2010 the Nature Conservancy sponsored workshops in collaboration with universities, federal government agencies, and NGOs. Employing National Center for Atmospheric Research methodology, these groups have found a more definitive, troublesome narrative. Their downscaled models posit two scenarios based on projections from the North American Regional Climate Change Assessment Program. The main scenario forecasts annual temperature in the Bear River Basin increasing by 6.3 degrees F, with annual precipitation growing by 1.6 percent (and that additional moisture falling as rain rather than snow). This scenario would lead to a 5–18 percent decrease in annual runoff into the Bear. Snow wouldn't begin to accumulate until later in the fall, with 10–15 percent lower peak accumulation. Spring snowmelt and runoff would occur up to three weeks earlier, with summer flows dropping 10–20 percent and winter flows increasing 30–50 percent due to more rain events. A second scenario suggests a 3 percent decrease in precipitation, a 4.9 degree F temperature increase, and a 5–13 percent decrease in river flow.[13] Gillies points out that these two scenarios probably are out of date because they didn't know in 2010 that the rate of carbon dioxide being dumped into the atmosphere is accelerating faster than modeled.[14]

The Bear River ecosystem will feel some of these effects of global warming harder than others. If water temperatures rise due to diminished river flows, especially in summer, warm water–adapted brown trout might move higher in their drainages, outcompeting the native Bonneville cutthroat trout, which need colder water to thrive. If the snowline moves higher in the basin, leaving lower elevations bare, and if more winter precipitation falls as rain in severe storms, it's likely that sheet runoff would carry more sedimentation and contaminants from adjacent fields and tributaries into the main stem of the Bear. Already drought is leading to less grazing forage in higher elevations of the Bear River Range. Because the number of cattle grazing on those pastures hasn't been reduced, dust bowls are developing. Less water in the Bear also means less water flowing into Great Salt Lake, imperiling its environmentally rich ecosystem.

Money, the Mother's Milk of Water Development

In his book *Cadillac Desert* Marc Reisner quipped that water can defy gravity and flow uphill to money.[15] What happens, though, when money begins to dry up?

Culinary water rates in Utah currently are subsidized by a combination of property taxes, state sales taxes that go into a water-development fund, and federal subsidies. So Salt Lake County residents, beneficiaries if Bear River water is diverted for their use, pay just less than 50 percent of the real cost of the water they use.[16] This subsidization encourages waste because residents don't pay out of pocket for all the water they use. Moreover, the property tax subsidy makes alternative water sources look more expensive than they really are, as their costs of delivery aren't masked by property tax subsidies.[17] When funding sources other than property taxes are suggested, such as higher user fees or hookup levies on real estate developers, the Jordan Valley Water Conservancy District and other water purveyors steadfastly argue that they need the reliable income from property taxes to float bonds for future development at low interest rates.

But the Utah State Legislature shows some signs of being willing to sever property taxes from water funding. The issue brings together odd bedfellows. The Utah Taxpayers Association feels that user fees would be more appropriate. The public education community would like to see more property taxes going to education. Sensing resistance to property tax subsidization, and trying to lessen the burden on Washington County taxpayers facing exploding impact fees for local water developments, state legislators from southern Utah are pressing to earmark 15 percent of future sales tax revenues to water development. If all state residents paid more into a water development pot, goes the reasoning, the astronomical costs for water development—$1.5 billion-plus for the Bear River Project and $1 billion-plus for the Lake Powell Pipeline, according to a 2012 Utah Legislative Research Council report—could be defrayed over a larger taxpayer base.[18]

Water development schemes can be as legendarily slippery as the claims found in Old West gold mining prospectuses. Southern Utah interests sense that the Lake Powell Pipeline would have a better chance of being partially funded through statewide sales taxes if the Bear River Project could be lumped together with it, because several million northern Utahns would benefit from Bear River water and the 141,000 residents of Washington County could sneak in under that shell. During its November 2012 meeting, though, the Revenue and Taxation Interim Committee voted down the request to increase the state sales tax earmarked for water development. For a state historically promoting the greening of the desert, that's a watershed moment.

Ironically, money might wind up reversing the historical course of water financing. Because water still needs to be paid for and bond holders want assurance, user fees loom. Economists maintain that tiered pricing, where consumers pay for the water they use, with incrementally higher unit prices for higher usage, is the best way to price the true value of water in the desert. With all levels of government having to deal with lower revenues because of the recession and economically strapped citizens less able to pay increased taxes, the spigot for water development might close to a slow drip. The era of dam building by the Bureau of Reclamation and the Army Corps of Engineers ended with Bill Clinton's administration. With even fewer dollars to spread to states, federal officials more carefully scrutinize water development proposals, looking for bloated population growth projections. Moreover, the effects of climate change will likely put more pressure on limited federal largesse. Faced with repairing damage from climate change-related storms like Hurricane Sandy, for instance, political pressure for reconstruction dollars from electorally powerful eastern seaboard states will trump water development pleas from Utahns hoping to be spared the real cost of developing desert water.

The Bear River Spring

Faced with the hard figures, water purveyors view conservation as the best way to create "new" water in the near future. The Jordan Valley Water Conservancy District believes that it can realize a 25 percent reduction in water use through conservation by 2025.[19] The Utah Rivers Council (URC), however, feels that this goal is too modest. It believes that the district can find another 10 percent savings, for an overall 35 percent reduction in water use by mid-century. The URC also has identified 60,000 acre-feet of legacy irrigation water in Salt Lake Valley that isn't being used, more than enough water to replace the 50,000 acre-feet from the Bear River Project.[20] But as the Jordan Valley Water Conservation District points out, that would be expensive water: it would require purchasing those water rights (if the sellers were willing) and then adding new treatment plants.

The state itself only wants to set modest goals and has targeted a statewide 25 percent reduction of water use via conservation by 2050.[21] Restraint is a challenging concept for a state that tends to favor exploitation over stewardship, even when self-discipline is a mantra of conservatism.

Still, other groups are moving ahead, trying to avoid stepping into the stagnant backwaters of political intransigence. The Jordan Valley Water Conservancy District has planted demonstration gardens to show Utahns how to make the desert bloom with less water. It also provides model water conservation landscape ordinances for commercial real estate developers and municipalities. Right now, however, the ordinances are only voluntary. Utah State University extension agents, in partnership with the JVWCD, offer free water audits for residents wondering if they are overwatering their lawns. Although the state's Slow the Flow public relations campaign is poorly funded, it at least makes conservation visible to television audiences.

In Cache Valley ranchers are building concrete bunkers to store manure so that it doesn't run into Bear River tributaries like the Little Bear. They are learning, too, to spread the manure on their fields at the right time of the season, so that sheet flooding doesn't carry it off frozen fields and into the river. All this is part of the Utah State University's outreach programs to educate farmers on best management practices. Professor Mesner and others have produced documents outlining how to monitor water quality with clear, actionable objectives. The Utah State Division of Water Resources is partnering with others to look at nonpoint pollution, the primary pollutant threat to the Bear River, via aerial photography and high frequency monitoring stations. Even though the state has limited financial resources due to budget cutbacks (and once opposed using outside funding for stream monitoring when it was available) it is now monitoring smaller watersheds, reasoning that smaller watersheds are more amenable to planning and field implementation.

Smaller governments that see the Bear River as a tourism resource, like Evanston, Wyoming, are pitching in, too. At one time the river was artificially channeled through town. Now Evanston has put back the Bear's natural meanders and has engineered structures like riprapping and instream jetties to prevent stream bank erosion and create alternating riffles and pools for trout habitat. Box Elder County brought ranching and conservation interests together to write a proactive Special Area Management Plan for the Bear River/Great Salt Lake wetlands to avoid protracted battles over mitigation resulting from helter-skelter urban development.

While dam building and hydropower generation have drained the spirit from some reaches of the Bear River, ironically, those same forces are tenuously bringing life back to the river. In 1999 PacifiCorp filed to relicense its hydropower operations at Soda Point, Grace/Cove, and Oneida reservoirs, and this regulatory process provided new interests with an opening to stake a claim on the river. Not accepting PacifiCorp's claims that the Bear River was a degraded environment, and that the company had minimally affected its already-poor condition, NGOs forced PacifiCorp to accept compromises that would lower hydropower production. The Idaho Council of Trout Unlimited, Idaho Rivers United, the Greater Yellowstone Coalition, and American Whitewater demanded that recreation and environmental mitigation be part of the new agreement. Fishers wanted PacifiCorp to recognize the need to figure preservation of the Bonneville cutthroat trout into its power-generating equations.

In settlement negotiations PacifiCorp recognized that ecosystem degradation was a problem and agreed to restore native fish habitat through mitigation funding and to conduct studies leading to the development of a Bonneville cutthroat restoration plan. Boaters won the concession to allow greater early summer flows through Black Canyon to sustain Class I and Class II whitewater opportunities. Perhaps most portentous, PacifiCorp agreed to maintain the health of the river and recreation opportunities from Oneida Reservoir downstream to Oneida Narrows, a distance of about seven miles. That set the scene for a clash between the Old West and New West.

Twin Lakes Canal Company had a problem. It couldn't store its Mink Creek water rights during winter because pipes carrying that water to downstream reservoirs that it owned would freeze. Twin Lakes proposed a dam at Oneida Narrows to store those water rights. It would pay for the dam by generating hydropower. Feeling reasonably confident because PacifiCorp had been doing the same thing for a century, it applied to the Idaho Department of Water Resources for a water permit and to the Federal Energy Regulatory Commission (FERC) for permission to construct hydropower facilities.

Twin Lakes expected opposition from the same NGOs that had pressured PacifiCorp during its relicensing negotiations. But it assumed that PacifiCorp would support its application: after all, they were in the same business, using the same Old West streambed. Even though it was uncomfortable intervening in an issue that aligned with its historical interests, PacifiCorp opposed the Twin Lakes application because it was concerned that the new dam would interfere with PacifiCorp's responsibility under its new license to proceed with mitigation and protection of recreational facilities downstream from Oneida Reservoir.

In 2012 the Idaho Department of Water Resources (IDWR) twice turned down the Twin Lakes application for a water permit, declaring: "The proposed project conflicts with the local public interest."[22] While recognizing a public interest in hydropower generation and irrigation, the IDWR said that the benefits of hydropower and the "relatively small addition of 5,000 acre feet of storage for occasional irrigation don't justify the permanent inundation of the Bear River" through the Narrows. The agency pointed out that maintaining main-stem habitat in Oneida Narrows was critical for the success of Bonneville cutthroat trout restoration. Ultimately, the director of the IDWR said the Bear River through Oneida Narrows has "unique recreational and wildlife values" that would be lost if Twin Lakes Canal Company were allowed to build a dam to store its water rights.[23] FERC won't grant a hydropower license without a water permit, so a new dam on the Bear at Oneida Narrows is likely a fading dream floating in the memory of the Old West.

Those two decisions, the PacifiCorp relicensing agreement and Twin Lakes' failure to secure a water permit, are stimulating a raft of new environmental initiatives on the Bear River. Bolstered by the Idaho Department of Water Resources' characterization of the unique recreational values of the Bear, the Bear River Watershed Council and the Oneida Narrows Organization are creating a campaign to designate the Oneida Narrows stretch of the river as a Wild and Scenic River under the recreation criterion.

But it's in the area of riparian and upland habitat restoration where the greater changes are taking place and where resources are being deployed by major players. Signatories to the 2003 relicensing agreement—PacifiCorp, state and federal government agencies, NGOs, conservation organizations, and Indian tribes—formed an Environmental Coordinating Committee (ECC) to coordinate mitigation on the Bear. The goal is to create public-private partnerships to rehabilitate a river that for too long has been neglected. Moreover, the partners recognize that climate change must be met with adaptive land management. Uncertainty is the watchword going forward. To meet that challenge, restoration partners are devising plans that stress redundancy and diversity. No one is sure how the effects of climate change will play out, so it's important that habitats be managed for *resilience*.

The ECC then looked to the Nature Conservancy and its Conservation Action Planning model, a biologically driven process that guided project teams to identify threats to the Bear River's flows and habitats and to coordinate and prioritize conservation strategies. In 2009 the coalition began scoring the health of the entire Bear River Basin, targeting riparian, wetland, aquatic, and grassland systems. Conservation of these targets would secure the ecology and biodiversity.

Once planners identified targets where intervention would offer the most likely success, the coalition agreed to follow a blueprint generated by the Conservation Action Plan. The Bear was divided into three segments: Upper (headwaters to the first Idaho border), Middle (largely the river in Idaho), and Lower (from the Idaho-Utah border to Great Salt Lake). As biologist Eve Davies, PacifiCorp's representative on the ECC says, "It's an elegant plan because it's easy to see."[24]

PacifiCorp is contributing $45 million and is focusing its efforts on protecting 8,000 acres on the Middle Bear. It already had agreed to retire its Cove hydropower generating plant under the relicensing agreement and to use that water to create ponds for brood stock and maintain stream flow and water quality for Bonneville cutthroat restoration.[25] PacifiCorp currently is planning fish-friendly diversions around the Georgetown Creek area to protect Bonneville cutthroat spawning habitat.

Always interested in the Bear River because of its relationship to Great Salt Lake, the Nature Conservancy (TNC) is working on the Lower Bear and incorporating it into its Wetlands Reserve Program. While working on a variety of issues, TNC is concentrating on controlling invasive weeds that lead to biodiversity loss by crowding out native plants.

The Nature Conservancy also is working with Trout Unlimited to reconnect Bonneville cutthroat waters. For instance, TNC has a conservation easement on the upper Little Bear River. By switching from flood irrigation to sprinklers, the easement conserves water for potential instream flow. But while water law doesn't allow for permanent reassignment of water rights to instream flow, it does allow temporary use of conservation water for instream flow if it's for the protection of a native fish. Because Trout Unlimited is in that business, it can lease that water from the Nature Conservancy and use it to help restore Bonneville cutthroat trout connectivity in the Bear River system. This is just part of Trout Unlimited's work on the Bear, where it has restored thirty-five fish passages and improved or restored connections in about 150 miles of streams in the watershed.

The U.S. Fish and Wildlife Service (USFWS) long has been a player on the Bear: it manages three national wildlife refuges on the river. But those refuges are limited in size and are imperiled because of development. So the USFWS is looking to purchase conservation easements from landowners along the Bear to maintain important habitat for fish and wildlife by ensuring that it won't be developed. Eventually the easements could lead to habitat improvement. The USFWS is prepared to spend up to $100 million for easements that would keep the land in agricultural production by drawing from the federal Land and Water Conservation Fund.[26]

A Snowy Alliance

Change is imminent for the Bear, if only because the river relies on desert water. Alterations could be positive or negative for the watershed, depending upon how the Byzantine equation of climate change, development, mitigation, water law politics, and the economy plays out. Any prognostications would require a supercomputer to analyze the possibilities.

But it's possible that the dynamics of global warming are creating a perfect storm for galvanizing change. To help us see beyond one or two generations. To help us change course.

At first glance it appears that water law freezes historical interests into place. Change is glacial at best. But when we consider how dramatically the pace of dam building has slowed, and the willingness of some Western communities to practice conservation, a thaw seems to be occurring. The cumulative effect of a slow-moving mix of challenges—prolonged drought, unsustainable population growth, even something as seemingly minor as the faint rumblings of resurrecting plans to build a dam at Honeyville—could crack the historical veneer.

Right now the Utah ski industry is promoting SkiLink, a scheme to connect Wasatch Front ski resorts via a tram. The proposal requires public land to be transferred to a foreign private interest. The deal would prevent the U.S. Forest Service from intervening. But Salt Lake City and Salt Lake County, local governments that zealously guard the Wasatch Front canyons that supply the metropolis with water, are opposing the transfer because it could degrade water quality. SkiLink and the land transfer proposal currently are on hold while stakeholders explore a comprehensive regional transportation plan.

Ironically, skiing interests are trying to promote "the Greatest Snow on Earth" when that resource is imperiled by forecasts of less snow and a contracted winter season. You would think that the ski industry would first want to protect the long-term viability of its moneymaking resource. If it looked closely at the dynamics of the Bear River and Great Salt Lake and how they affect the amount of snow that falls on Wasatch Front resorts, it might redirect some of its considerable economic and political resources to conservation rather than development.

Consider how the lake banding effect affects snowfall in the Wasatch Front. Cold fronts from the northwest passing over the warmer water of Great Salt Lake pick up moisture and deposit it as snow in the mountains. Up to 10 percent of the snowfall in the Wasatch Front can come from this lake effect.[27]

The Bear River supplies 60 percent of the inflow into Great Salt Lake, but the Utah Rivers Council has determined that the Bear River Development Act would eliminate 18 percent of the Bear's flow into the lake in an average year and 70 percent during a severe drought.[28] Because it is a relatively shallow lake, any diminished inflow dramatically decreases its surface acreage. Now, consider that a 20 percent diversion of annual Bear River flow into Great Salt Lake would lower the lake two to four feet with a correspondingly dramatic drop in surface acreage. With less water already flowing in the Bear because of global warming, that would create two related challenges. The rich ecosystem of the Bear River/Great Salt Lake nexus would be deprived of its life-sustaining elixir. And a smaller surface area would mean less moisture from lake effect, giving ski resorts even less snow to promote. The view from SkiLink in glossy promotions won't look so appealing with patches of mud on the ground.

Fortunately, all water interests—developers, governments, conservationists, and water-quality watchdogs—recognize that conservation is the first and least expensive step toward meeting future water demands. With more widespread tier pricing and less reliance on property taxes, conservation would look even more attractive. Most important, conservation gives us time.

We need time to shed greening-of-the-desert myths that imprison our future. Time to reconsider the effects that diverting water from the Bear would have—on Great Salt Lake marshlands and our snowpack. Time to learn the lesson of what it means to rely on a river that begins and ends in our neighborhood. Time to change course.

Notes

1. "Bear River Development," State of Utah, Department of Natural Resources, Utah Division of Water Resources, August 2000, http://www.water.utah.gov/brochures/brdev.pdf.
2. Wallace N. Jibson, *History of the Bear River Compact.*
3. Bear River Commission, *Bear River Compact,* 1.
4. For a graph illustrating per capita water use by city, see figure 6.1 in chapter 6 by Zachary Frankel herein.
5. *Bear River Development Act,* Utah State Legislature, 1991; for details see *Bear River Development,* 8.
6. Ibid.
7. David Ovard, general manager Jordan Valley Water Conservancy District, personal interview, January 16, 2003; chapter 6 by Zachary Frankel herein disputes such claims.
8. Fred Selman, personal interview by Craig Denton, April 22, 2002; Hansen Park public meeting, Elwood, Utah, April 8, 2000.
9. Utah State Senate Bill 92, 2002, http://le.utah.gov/~2002/htmdoc/sbillhtm/SB0092.htm.
10. Nancy Mesner, personal communication, November 26, 2012.
11. Rob Gillies, personal communication, December 6, 2012.
12. Utah Rivers Council, *Crossroads Utah: Utah's Climate Future,* 11.
13. Joan Degiorgio et al., "Southwest Climate Change Initiative: Bear River Climate Change Adaptation Workshop Summary," Salt Lake City: Nature Conservancy, 2010, http://nmconservation.org/dl/SWCCI-BearRiver-Climate-Adaptation-Wkshp-FINAL-Report-Nov-2010.pdf.
14. Gillies, personal communication, 2014; for a further discussion of climate change in Utah see chapter 5 by Daniel Bedford herein.
15. Marc Reisner, *Cadillac Desert: The American West and Its Disappearing Water.*
16. *Mirage in the Desert: Property Tax Subsidies for Water,* Salt Lake City: Utah Rivers Council, 2002; see also Gail Blattenberger, associate professor emerita of economics at the University of Utah, "Property Tax and Water in Salt Lake County," white paper in author's personal possession.
17. For more discussion of the relationship between water waste and property tax, see chapter 6 by Zachary Frankel and chapter 13 by Daniel McCool herein.
18. *Utah State Legislative Research Council Report,* Summer 2012.
19. *Annual Report,* Jordan Valley Water Conservancy District, 2001.
20. *Waterlines* 28 (Autumn 2005), Utah Rivers Council, Salt Lake City, Utah.
21. Zachary Frankel, personal interview by Craig Denton, January 25, 2000.

22. Idaho Department of Water Resources, *In the Matter of Application for Permit No. 13-7697 in the Name of Twin Lakes Canal Co., Final Order Denying Application for Permit* October, 2012, 50.
23. Ibid., 49.
24. Eve Davies, personal communication, November 28, 2012.
25. Ibid.
26. Ibid.
27. Dr. James Steenburgh, personal interview by Craig Denton, March 31, 2006.
28. *Waterlines* (2005).

4

The Restoration of All Things

The Case of the Provo River Delta

GEORGE HANDLEY

Because they descend from considerable heights over relatively short distances, mountain rivers run loud, cold, and violent. They and the life they support evolved over millennia according to mountains' dramatic and short term changes in snowfall, snowmelt, and temperature. Speaking in terms of deep time and the interdependent webs of ecology that rivers represent, it is hard to say where such peripatetic bodies begin, where they end, or what their proper terrain is. The Provo River is no exception. It emerges from runoff and springs lying above 10,000 feet in the southwestern quadrant of the Uinta Mountains. Gravity pulls it down over some seventy miles as the crow flies, gathering force and building alliances with cousin tributaries in the watershed until it arrives at the delta at Utah Lake, at just less than 4,500 feet. After the arrival of Europeans a little over 150 years ago, this water was gradually harnessed, tamed, and ultimately subdued by irrigation, dams, extensive canals and is now threatened by the effects of encroaching development and climate change. The changes it has seen in its most recent history have proven so much more dramatic and sudden than normal that its normally resilient capacity for adaptation has proven unable to forestall modernity's devastating effects of degradation and biodiversity loss.

What is perhaps most remarkable about this now almost universal story of rivers is how little it seems to concern those whose lives and livelihoods most directly depend on rivers. Part of this can be explained, I believe, by the shallow memory of contemporary populations. It is hard to see a river's history. It takes real work—the work of historical reconstruction and imagination—to see in the mind's eye a mountain river of two hundred years ago, braiding and weaving through a wetland and supplying habitat for a wide diversity of life. Only with such a point of comparison in mind can we see just how damaged our waters are. The Provo flows peacefully through Utah valley and looks and acts like a reasonably healthy body of water except for the occasional milk jug or other piece of garbage caught in its bed. On its diked and channeled shores lie parking lots, shopping malls, and homes but also trails, parks, and softball fields, all suggesting suburban respite. The river is so utterly normal, at least to eyes that have grown accustomed to a world that we have made after our own image. But to comprehend nature's independence from what we have made of it requires eyes more attuned to enchantment and wonder, more welcoming of nature's surprising difference.

In 2004 my colleague Lance Larsen and I found ourselves across the table from the novelist Marilynne Robinson, who had agreed to an interview while visiting the campus of Brigham Young University. Her novel *Housekeeping*, about her home in northern Idaho, exemplifies as well as any novel I know how great art emerges from this kind of awesome comprehension of a home landscape. Lance asked her about the degree to which landscape shapes our self-understanding. She answered: "I'm beginning to wonder if I could make a distinction between character and landscape." Lance pressed her further, asking which comes first, human self-understanding or our awareness of our surroundings. She gave this stunning response:

> There's probably nothing stranger than the fact that we exist on a planet. Very odd. Who does not feel the oddness of this? I mean, stop and think about where we actually are in the larger sense. It seems to me as if every local landscape is a version of the cosmic mystery, that it is very strange that we're here, and that it is very strange that we are what we are. In a certain sense the mystery of the physical reality of the human being is expressed in any individual case by the mystery of a present landscape. The landscape is ours in the sense that it is the landscape that we query. So, we're created in the fact of ourselves answering to a particular sense of amazement.[1]

Robinson stresses that culturally speaking there is nothing natural about nature. It is the oddest and rawest of facts about our human existence that we would find ourselves anywhere in particular, confronted by nothing but the physical particulars of a place. Because it gives birth to a sense of amazement, which is also an essential form of humility, our home environment provides us with a vital opportunity to shape ourselves—our language, our culture, and our identities—in intimate relation to our own biology, rather than in indifference or hostility toward it. For Robinson, cultivating a greater sense of awe for the ordinary fact of the world around us is cultural work at its most ethical. Why, given the stunning beauty of a place like the Wasatch Front, do so many of us *not* feel this amazement? We certainly seem to have been in a hurry to engineer ourselves out of this state of stupefaction before the cosmic mystery, as if we wished life could be of our own making, as if the strangeness of finding us here on an earth of unimaginable age and of numberless forms somehow diminishes or even terrifies us. Because when we really consider the miracle of life—the odd chance of us being here and the wonder of so much extravagant expense and all for what?—we are reminded of our mortality and our fragile belonging in an ancient home of mutual interdependence. The paradox is that in this realization of our relative insignificance we also discover our unique human privilege: the privilege to experience wonder in the face of nature's variety, beauty, and grace. There are several reasons to believe in our exceptionality in the creation, but surely our capacity for awe must be one of them. As the theologian William Brown puts it, perhaps we should rename ourselves: not so much *Homo sapiens* (knowing man) as *Homo admirans* (wondering man).[2] So it is a particularly sad waste of human capacity to squander the chance to feel gratitude for the gift of being alive.

The landscape is just one temporary and particular manifestation of a life force that for millions of years has shaped mountains and rivers, valleys and lakes, animal and plant forms. And all of these forms are a creative response to the particular limits imposed by time, weather, and place that physical life imposes on all. To have life, to have existence as human beings, is to have inherited relationships among particular plants and animals, particular watersheds and land forms, and particular climates where we find ourselves. These are relationships born of improvisation, collaboration, and imagination. Our charge, it would seem, is to learn the particulars of where we live and live creatively in the interest of the ongoing health and flourishing of all life forms. When we contemplate the temporal and spatial range of creation, we can only conclude that to live and die for human flourishing alone is puny and insufficient gratitude for life, especially when human well-being is so often defined by short-term materialism and comes unnecessarily at the expense of the rest of life. Pessimists will say that we have to choose between the proverbial logger and the spotted owl. My LDS faith suggests otherwise. It teaches that all forms of life—human and more-than-human—consist of both body and spirit, that plants, animals, and humans alike are "living souls" that have the right to joy in posterity, a providentially sanctioned purpose to flourish. Why else would God in Genesis 1:22 have commanded the fishes to fill the seas and the fowls to fill the waters? Why would he deem the waters' capacity to bring forth life "abundantly" a "good"? To multiply and replenish is a mandate for all life, and LDS creation theology only provides more reasons to believe that our unique human challenge, as God sought so hard to teach Job, is to understand the importance of our human experience and sufferings in the context of broad geography and deep time, as a central part of the immense fabric of endlessly diverse life.

This passage from Genesis notwithstanding, Christians have not always been keen on addressing their responsibility to assure the health and well-being of the whole system of life, preferring instead to hear God's commands as giving humankind special license to do with nature as we please, to have dominion as lords rather than accountability as stewards. As we have gained increasing levels of control over natural resources, many Christian theologies imagined that our stewardship is precisely to learn to liberate ourselves from the physical restraints that we experienced for millennia in the natural world. This kind of Christianity translated into a politics of "improvement" of nature, always seeing untouched nature as fallen and in need of the redeeming effects of human engineering. These are theologies that see our embodiment and our emplacement in specific geographies and ecosystems as temporary limits and maybe even enemies to our spiritual destiny. Even as evidence mounts that the environment is suffering at our hand, the evidence is either denied or embraced as a sign that the prophesied end is near.

As many eco-theologians have argued, this disparaging of the body and of the deeply temporal nature of evolved life was fueled by a refusal to grant to nonhuman life any degree of spiritual or intrinsic worth, let alone any kind of intelligence. Human exceptionality rose to such a height of overreach that we failed to see any longer our strange and marvelous and humbling kinship with all living things. We lost, in other words, our capacity for wonder, except perhaps wonder at our own achievements. The result, of course, has been that Christianity, despite its rather ecologically friendly ethos, saw technology and the exploitation of nature for human ends as a necessary means to advance spiritual goals.

As thematized by historians such as Wallace Stegner, Marc Reisner, and Donald Worster, Mormons, of course, have played no small role in the engineering of the American West. They sought to make the desert blossom as a rose, not necessarily through supernatural, aesthetic, or metaphorical means but through increasingly advanced technologies of irrigation, including the straightening and damming of rivers over the course of the past century. Despite a rich history of stewardship, significant environmental commitments by the Church of Jesus Christ of Latter-day Saints, and the work of many LDS environmentalists, the political culture of Utah is more known for its history of opposing environmentalism than for any particularly strong record of effective and sustainable stewardship.[3] Moreover, despite its reputation as an ecologically unfriendly theology, LDS belief about the creation strongly resembles the work of many of the world's leading eco-theologians. Mormonism's many earth-friendly beliefs need greater attention if we are to hope for a change in Utah's political culture and for a more pronounced Mormon ethic of concern for the ecological crisis.

It has taken the better part of the past fifty or sixty years for the science of ecology to develop to the point where we can now understand our relationship to water in evolutionary and ecological context and therefore more clearly comprehend the damage we have done to the diversity of life that fresh water uniquely fosters. And the science now provides us the opportunity to work toward the restoration of ecosystems. These restoration efforts are not perfect and are sometimes criticized because they involve a significant degree of violent intervention into damaged ecosystems in order to restore better ecological health. Moreover, it is simply not possible to know in the relatively short term if restoration projects are worth the risk and expense. The solution, however, is not excess caution in protecting an unacceptable status quo; what we need is a stronger long-term ethics.

My hope is that ecological restoration might take root as an ethic within Mormonism. I find the LDS Church's effort to restore the sacred grove where Joseph Smith received his first vision a promising start. The word "restoration," of course, has particular resonance in a Mormon context. Jesus spoke of a time when Elias would come and "restore all things." As LDS theology has it, however, this restoration is not a one-time event but an ongoing and continual unveiling of what once was. Mormons believe that Joseph Smith inaugurated this prophesied "restoration of all things"; along with saving ordinances and doctrines, this restoration requires the great work of gathering the broken body of all truth into one. Restoration is the fruit of revelation, but it is also the work of research, of probing questions, of fact-finding, and of dialogue with the world; it is an unearthing what was once known and discovering what is yet to be known. Restoration theology, in other words, is hopeful, especially about knowledge, but it is also a reminder that knowledge is always contingent upon our willingness to do hard intellectual work, research, rethinking, and, I suppose, repenting.[4] We go back, as the word "repentance" implies, in order to go forward. To restore requires a proper awakening, a refilling, of our spiritual capacities through a recovery of knowledge.

As a Mormon environmentalist, I like the idea that restoration includes a recovery of an understanding of workings of the world that science discovers every day. I like to think that scientific knowledge is really just ancient understanding, something that in some deep sense we have always known, because it is knowledge that might bring the world back into its native ecological health. Indeed, among Joseph Smith's remarkable teachings, surely his creation theology—which emphasizes the spiritual qualities of all of life and the centrality of the earth in our understanding of the meaning of heaven—must factor in as a major contribution to our need to return to deeper knowledge of and reverence for the earth's capacity to generate life in all of its forms.

If Marilynne Robinson is right, we bear a responsibility to create the meaning of our existence not out of the substance of things—things that we can make out of the flesh of the earth or things that we can own—but out of the substance of our amazement. One wonders what our economy and our culture would look like if we lived with a deeper appreciation for the mystery of physical existence. It would certainly mean less interest in material possessions and deeper respect for our places of inhabitation.

Robinson says that the landscape is not ours in the sense that we own, engineer, or make it after our own image. It is only ours to the extent that we detach ourselves from proprietary notions of permanence, invulnerability, and reliable security. To be amazed is to be astonished. Thunderstruck. Dumbstruck. Unable to speak but lit up, connected to heaven by an electricity that destroys shallow and self-absorbed individuality and, in its place, plants the seed of recognition of our great belonging to all things. Then maybe we learn that we don't have to choose between cherishing our differences from other life forms or being dissolved by our commonalities.

Robinson seems to imply, and many eco-critics concur, that poetry, music, and art are often born out of a desire to strike a more suitable balance. The arts are often the recuperation of speech after speechlessness, a discovery of self in the midst of expansive wonder. And unlike the vast treasures that we have taken so thoughtlessly from the earth, these are the clean and renewable resources of astonishment. Renewable and *renewing* resources. I do not mean merely our love for Ansel Adams calendars or our love for the exotic and extreme forms of beauty that we have photographed, painted, and praised as a nation in the Tetons, Yellowstone, the Grand Canyon, or our beloved Zion. Love of exotic and extreme natural beauty that does not stem from an intimate love for our local landscape is the equivalent of pornographic desire. It wants nature as a thing that pleases when we pay and that we can ignore at will, but not as a partner in a committed relationship of mutual and unconditional flourishing.

Maybe we resist the task of caring for nature here at home because everywhere we look we see memorials of our own clumsy and often thoughtless transformations of the wilderness. The health of my home waters—especially the environmental health of the Provo River, Utah Lake, and the air we breathe in the valley—has been severely compromised. What was once a thriving watershed that provided life on the valley floor for a broad diversity of plants, fish, large and small mammals, birds, insects, and human populations for thousands of years has, in a mere century and a half, become a narrow channel of water through a wilderness of concrete. This has chased the animals to higher ground and driven eleven species of fish, native only to Utah Lake, to extinction. Only two native species of the lake have survived this war. One, the June sucker, is the keystone species of the watershed but is endangered, even though it is perhaps the single most important reason why our pioneer ancestors managed to avoid starvation in this valley. A neighbor of mine tells me that his pioneer grandmother who homesteaded in Provo told him of those days. She at least felt the need to tell her descendants to honor that gift through stewardship of the river and lake. One would think that we in Utah Valley would have felt dismay at this impending loss a little more keenly.

Some do, however, and what is especially encouraging is that no small number of them are Mormons. A proposal has emerged to restore the Provo River delta and the sucker's spawning ground to its original home just north of its current location on the shores of Utah Lake. This is no easy sell, especially if you listen to the anger of our citizens over the proposal. Pessimists call restoration hopeless nostalgia for a past that we cannot recover, but ironically, like addicts, they seem obsessed with going back again and again to the same mistakes of the past. And like addicts, they never explain why we should believe that doing nothing to mitigate our errors will lead us out into a brighter future. Restoration is messy and hard work, but like repentance, it is turned to the future and not the past and earns us a closer and more intimate appreciation for where we live. What restoration makes of the land is significant also for what it makes of us.

The story of the proposed restoration is as complex as the ecosystem that it hopes to help. It brings together various strands of the political culture of Utah and various economic interests, and to a large degree it has pitted Mormons against Mormons; it is not a stretch to argue that the resolution of the issue may depend on which strands of environmental belief within Mormonism will win out. As I have suggested, Mormonism is rife with some of the most earth-friendly doctrines in the Judeo-Christian tradition, but it is also a complex worldview that has at times led to competing visions about economic values and environmental ethics. It might be useful to hear these competing ideas fleshed out in their more polarized manifestations. I do not mean to suggest that these positions represent anything official.[5] Indeed, these are more like folk theologies informally or implicitly invoked by those on both sides of environmental issues.

When LDS theology is used as a justification for anti-environmental attitudes, it goes something like this:

1. The Parable of the Talents: One often hears this parable cited informally in LDS circles as evidence that the Lord is never pleased when we are given resources and choose only to preserve or protect them

from harm. To worry unduly about conservation is to adopt fear and exhibit mistrust in ourselves and in God. This fear, it is believed, is not warranted by the gospel of hope and eternal progression that are cornerstones of LDS belief. True stewardship, then, means that we should make full and creative use of God-given resources and make improvements on a fallen world; we should, in short, develop or "dress the Garden." Improvement of the earth is a virtue and, by implication, undeveloped nature is fallen, in need of human-engineered redemption. The implication has been that what we can make of the world is a test of our creativity and our trust in the spark of divinity within us. Some argue today that to bypass the use of such things as fossil fuels is to ignore that they have provided for the improvement of the human condition under modernity and the meteoric rise of the church in the last days. They were God-given, in other words. It would be ingratitude to God to imagine the use of such resources as a mistake.

2. We Are God's Spiritually Begotten Children: LDS belief is arguably one of the most anthropocentric (human-centered) theologies in Christianity because it elevates human significance to the status of being "gods in embryo." Human beings are offspring of divinity, and this fact alone makes us stand in stark contrast to every living thing around us. This idea contrasts virtually all environmental philosophies, all of which imply if not directly state that the root of our environmental indifference and poor caretaking of the natural world is an inflated and overstated sense of human significance. For many LDS, worry about the environment is a symptom of a failure to understand the spark of divinity with us.
3. Eternal Perspective: Because LDS theology stares so deeply into eternity, the implication is that LDS believers should expect to live with a keen sense of the temporary nature of earthly conditions. If we understand our bodies, this earth, and all of creation as the rhetorical stage upon which we are learning to work out our eternal progression, then what happens to the earth in the process of doing such work is merely part of the plan and is certainly accounted for by divine providence. The idea is that we will make mistakes, to be sure, but we ought not panic and should stay focused in our trust in God's purposes. What is lasting and eternal is what becomes of us, not the conditions of this earth.

4. Family Comes First: The LDS Church is a pro-family religion, one that believes that procreation is itself divinely mandated. Human beings are to multiply and replenish the earth, and philosophies that denigrate well-intended parental desires to have a family or erode the freedom of parents to decide the size of their own family are to be avoided. Moreover, the family is the locus of our most important moral and ethical considerations. If we adopt philosophies that point us beyond the home and beyond concerns for human well-being, we might run the risk of compromising our ability to sustain family life.
5. Freedom: Free will is central to LDS theology. For many LDS, the perception is that we have seen a gradual erosion of freedoms over the course of the last century, particularly as a result of secularism. Environmentalism often requires large-scale solutions that impinge upon the freedoms of individuals and communities and is therefore to be treated with suspicion.
6. The End Is Near: A church that has "Latter-day Saints" in its title is unapologetically focused on the final preparations for the Second Coming. To worry about sustainability or long-term environmental effects of our actions is to miss the point. We should not only be focused on eternity but also on the relatively brief future ahead of us before we enter into the millennium where much of these concerns will be taken care of.

When Mormonism is used as a justification for pro-environmental attitudes, in contrast, it goes something like this:

1. The Spiritual Creation: All living things—plants, animals, and human beings alike—are "living souls" and are dignified with the opportunity of mortal experience and its chances for happiness. Happiness comes in the fulfillment of purpose, in the flourishing and reproduction of all life. This is taught in the LDS account of the creation. Everything, therefore, also has intrinsic worth and has a right to enjoy posterity.
2. Law of Consecration: Yes, LDS doctrine teaches that "there is enough and to spare" and that marriage and childbearing should be encouraged, but there is no mandate to have as many children as possible but only as many as a husband and wife decide to have together in an equal partnership. The verse from Doctrine & Covenants 104, from which

this phrase is excerpted, only guarantees that the earth will sustain the world's population if we overcome economic inequality, presumably through more equitable distribution of the earth's resources. Indeed, if we are to protect the sacred nature of the family and the central importance of reproduction in marriage, many Mormons feel that we must be equally serious about the call to live modestly, to redress poverty, and to consider future generations.

3. We Are Nothing: This is a companion doctrine to the notion that human beings are distinguished by being spiritually begotten children of God. In Joseph Smith's restored account of the creation in Genesis, Moses has a vision of the creation not only of this world but of the many worlds under God's care. He is overwhelmed by what he has seen, collapses, and says, "[N]ow for this cause I know that man is nothing, which thing I never had supposed" (Moses 1:10). This moment is akin to Job's humbling experience in listening to the Lord's description of his myriad creations. Both men learn and have faith in the fact that they are sons of God, of divine parentage, but they also must learn requisite humility, lest this idea of their own spark of divinity become an excuse to see their own importance out of proportion with all of life. Human self-importance must be tempered by the radical humility that comes from a true apprehension and appreciation for the wide and expansive diversity of God's creations. Our happiness is the central purpose of the creation, but this must not be used as an excuse for selfishness or disregard for God's love of his creations. Our divine parentage is a call to service, as Mormon thinker Hugh Nibley once said, not a license to exterminate.[6]
4. Freedom: For some, the need to protect our freedoms is a reason to be suspicious of environmentalism. For others, it is a reason to be concerned about environmental degradation. We are not free when we can't choose the quality of the air we breathe or the water or food we ingest. Pollution, waste, toxins, and climate change all impact human health and make us increasingly less free to live our lives. We are often not in a truly free market, able to choose sources of energy, methods of transportation, clothing, and food that are calculated to be gentler on the earth. We are seduced by a commercial society, slaves to our own greed. Environmentalism is one way to fight our way back to greater freedom.

5. Aesthetic and Spiritual Pleasure: All things are made for human benefit, but chief among those benefits appears to be aesthetic and spiritual pleasure. Nature is not intended to clothe us and feed us alone. It is also to "please the eye and to gladden the heart…to strengthen the body and to enliven the soul" (D&C 59: 18, 19). Adam and Eve were commanded to "dress" the garden but also to "keep" it. In short, the world is a test of our ability to distinguish between needs and wants, to find deeper and more sustainable pleasures in accepting the bounties of the world independent of what we might need or want to do with the earth's resources. We are not to be wasteful or excessive or be content to live at a material level above others.
6. The End Is Near: The feeling that the end is near, for some, is precisely what motivates stewardship. Knowing that our time here is limited only makes what we experience here all the more sweet, and knowing that providence provides assurance of an end that will not be tragic but redemptive inspires a desire to be worthy of such a magnificent gift. Worthiness can only come by working to make this world, here and now, more like the world to come.

I have engaged in this rhetorical exercise in order to tease out the reasons why Mormons who share the same beliefs might nevertheless have very different environmental attitudes. I don't pretend to have captured the logic definitively in an otherwise rich and diverse theological tradition. However, one thing is clear: if a watershed like the Provo is to stand any chance of recovering from the damage done to it over the past 150 years, Mormons need to find deeper motivations for taking care of the planet. I say this because they constitute more than 80 percent of the population of Utah Valley, where the Provo River flows. This will involve learning to strike a balance between these tendencies. Mormonism asks us to live with a deep appreciation of our central human significance but to do so with profound humility; it asks us to accept the earth's bounty as God's gift to us for our well-being and to work hard for our own provision but to live in such a way to provide for the poor and for all of life's flourishing;[7] it asks us to understand the importance of family and children without losing sight of the right of all living things to posterity; it asks us to be creative engineers of our constructed world but to appreciate the beauties of an untouched "unimproved" nature; it asks us to protect our agency but to learn to work together with others so that freedom is never merely individual but also social; it asks us to keep our eye on the eternities while seeing the life we make here and now as the most important manifestation of our spirituality.

This is really not so hard as contemporary rhetoric about environmentalism makes it sound. The fact is, political ideologies think more narrowly on these topics and try to present false choices for their adherents. So while religion is often perceived to be at the root of our environmental problems, it is not religion per se but political ideology that has run interference on our ability to think more clearly and deeply about the meaning of belief. To solve the ecological crisis, more religion, more research, more repentance, and more rethinking are necessary, not less.

A restoration of a delta like the Provo's would represent a victory of stewardship principles and of a collective ethos over individualistic, religious, and corporate short-termism. The facts are these: the delta of the Provo River, as it currently exists, is a ruin of what it once was, a meandering, braided wetland that—given its warm and shallow waters, varying and strong current speeds, and broad spread across the land,—sustained a wide variety of invertebrate life and provided a home for up to thirteen species of fish that were native to the lake, myriad bird species, and many large land mammals.[8] After generations of river straightening and dykes that began in the 1920s as a means of flood control and due to effluence and dams small and eventually large, the river delta found itself constricted into its present shape, unable to wag its tail in deep time up and down the northern slope of Provo Bay, according to changing geographical conditions, differing levels of sedimentation and drifting detritus from the 70-mile journey down from the Uintas, as it had over for millennia. Through careful study of soil, geography, and satellite images, we can now see where this delta moved and shifted back and forth over thousands of years, providing a fertile womb for diverse life. It now enters the lake on the southernmost tip of this ancient delta and only in one slow and riprapped line, a veritable bathtub of cold, sparsely vegetated water. To the north lies dried land, still pumped to this day to maintain it dry, owned by ranchers, while the almost still waters of the compromised delta provide little or no cover for the June sucker. The June sucker sustained human life on the shores of the lake for thousands of years and had been harvested heavily by the Ute Indians in recent generations when the

Mormons arrived. As I mentioned, as one of the most reliable and ample supplies of protein in the valley, the June sucker saved the Mormons from starvation.

But by 1986, when the June sucker was listed as an endangered species, it was no longer successfully spawning up the river. Its larval fish need food, vegetation, protection, warmth, and a strong current but could only find the dead zone of the delta, where the river lies up to fifteen feet deep (the lake, by contrast, averages six feet in depth), where the waters grow cold and anoxic, and where motorized boat use adds to the difficulties. What's worse, carp were imported in the early twentieth century as a commercial meat source, and now millions of carp range throughout the lake and dominate every corner of its ecology. The June sucker is an exceptionally well adapted fish of long vitality, with a life expectancy of up to forty years. Unlike trout, which are built to spawn yearly, the sucker has evolved to spawn more eggs and to spawn intermittently, thus allowing it to survive the dynamic and ever changing circumstances of mountain rivers, subject as they are to the whims of precipitation and weather and the likelihood of unsuccessful or less than ideal spawning conditions. Its long life is also the reason why it had survived, at least temporarily, the loss of a habitat needed for its reproduction. Like a great many endangered species, had it not been listed, it would not have survived such rapid and unprecedented human-caused change.

The Endangered Species Act, of course, is not a very popular law in Utah and the June sucker designation was not greeted with universal gratitude. The force of the ESA is not so much in the funding it provides for mitigation and restoration. Indeed, given its already stretched resources, the ESA only provides a modest $50,000 a year to protect the June sucker. Its force comes more from the threat it poses to current water projects under the auspices of the Central Utah Project (CUP). Utah Lake is the hub of CUP's operations and Fish and Wildlife Services has asked that more be done to protect the June sucker. Otherwise CUP might find itself themselves shorthanded with the resources that it needs for various projects. State agencies have wanted to avoid such a showdown between the state and the federal government, and this pressure has created a unique collaboration between the State of Utah and the Department of Natural Resources, Central Utah Water, Utah Reclamation Mitigation and Conservation Commission, Bureau of Reclamation, Fish and Wildlife Services, and two local users: Provo River Water Users Association and Provo Reservoir Canal Company. These organizations have been able to pool sufficient state funds to create the June Sucker Recovery Implementation Program.[9] Contrary to popular perceptions among Mormons that tend to see environmentalism as federally mandated and largely propagated by non-Mormons, such conservationism is a state—not a federal—effort shaped in the majority by Mormon players. I mention this because by now it should be a tired and worn-out stereotype that Mormons only want to dam rivers or that they don't care about ecological health.

It is fair to wonder, of course, if such a collaboration is sustainable and if it is guided by a sincere desire and commitment to bring the June sucker back from the brink of extinction. In other words, is the June sucker recovery an effort of convenience, even desperation, or is it an effort of principle? How much of a religiously grounded environmental ethos is motivating the work in a proactive fashion and how much of it is fear of federal intervention? The answer to this question matters because it tells us something about the prospect of a mainstream Mormon environmentalism. No doubt fear of federal intervention loomed large, especially in the beginning, but in my conversations with a number of LDS biologists and water specialists who work on this project I saw a palpable, genuine, and altruistic commitment to make good by the river and the sucker. For them, it feels like doing the right thing. They feel a sense of moral accomplishment in being able to reduce inefficiencies in water distribution and thereby provide a sufficient flow in the river year round. The excitement surrounding the first successful spawn of the fish in the restored delta of Hobble Creek, an initial restoration project that began on a smaller scale just a few miles south of the Provo, seems to have converted all but the most skeptical. Within a matter of a few months of the restoration, the June sucker, guided no doubt by its ancient instincts, was spawning upstream, apparently using all of the tributaries of Hobble, and fingerlings have managed to survive the winters in some of the pools and backwaters. Over the last four years the dominant invasive species of reed that grows along the lake shores known as phragmites has been defeated along the Hobble's delta. These efforts seem to have purchased enough goodwill in Washington to provide sufficient support in the years to come, if at this point the scientists can convince the landowners to sell enough of the delta's range for a meaningful restoration to take place.

This is a big if. The money is likely to be available to buy the land when the time comes. Some of the families, especially those who have come into the land more recently, want to sell, but it is the more longtime owners—those who have enjoyed a long history working the land and who are struggling to resist encroaching development—who are having the hardest time conceiving of parting with the land. And depending, of course, on who sells and on what sells, the restoration of the delta is likely to be seriously compromised. And without a successful restoration of the delta, the long-term health of Utah Lake is at risk. We can restore a delta that is smaller than its ancient size. But given prolonged droughts, reduction in snowpack, warmer springs, and other consequences of climate change that we are seeing, we are already facing a possible reduction of water flow of up to 50 percent in the coming decades. River restoration projects mitigate against the effects of climate change, but they also need sufficient space to be done properly. The good news is that landowners are showing signs of being willing to cooperate in the joint interest of restoring the delta and shoring up protections against suburban sprawl. Such a development was previously hard to imagine in a state where federal environmental regulations have typically had a bad name. With regard to another and perhaps more controversial listing, Representative Jason Chaffetz, for example, said about the sage grouse: "The only good place for a sage grouse to be listed is on the menu of a French bistro. It does not deserve federal protection, period."[10] This kind of rhetoric, of course, gets votes in Utah. It is not uncommon, however, for the very same politicians quietly to offer support for initiatives such as the June Sucker Recovery Project because no one is interested in making the CUP an endangered species. The sad truth is that, despite the richness of LDS theology, a respectable history of good stewardship, and the many Mormons who care deeply about the health of this valley precisely *because* they are Mormons, the population mostly plays willing hostage to the ideological fads and rhetoric of partisan posturing.

This in turn creates the mistaken impression that environmental concerns are largely the domain of non-LDS liberals. Indeed, the "intra-Mormon" character of this struggle, if you will, is largely lost on most folks in Utah. Even in Utah County, this struggle is often portrayed along the partisan lines of the rights of local families threatened by giant federal government bureaucracies, government overreach, or Gentile easterners telling us how to use our resources. A simple encounter with the players involved with the June Sucker Recovery Project, however, provides ample evidence of a strong Mormon and local contingent who have devoted their lives to stewardship of the river and the fish, who are graduates of Utah universities, and who have a determination to keep the federal government out of their business.

As someone already convinced of the value of the delta's restoration, I must confess that I struggled to sympathize with the initial hesitancy of the landowners to sell. But as I listened to their concerns, I could understand their attachment to the land, to family tradition, and to a place they wish to preserve, a fact which only makes their willingness to compromise that much more admirable. Besides, the loss of rural space in the valley already has a long and tragic history. Indeed, until the restoration project emerged, their land was already losing ground to the greed of developers and the indifference of citizens. But it is also true that Utah's proud culture of defensive stances against outsider intrusion is sometimes motivated by an overdeveloped territorial and proprietary conception of land and resources, to the point that we would allow attachment to neighborhood or to cultural traditions to override important and needed ecological and collective benefits. I used to agree with Wallace Stegner and Wendell Berry that indifference to place was a root cause of environmental degradation, but the history and experience of Mormons in Utah would suggest otherwise. There are more degraded places in the world, to be sure, and Mormonism has had its moments of exemplary environmental stewardship. Indeed, I cannot agree with the caricature of Mormons as willful and indifferent exploiters, but one still has to wonder why Mormon stewardship has not yet managed to stand out as particularly far-sighted or especially ecologically grounded, especially given a fierce Mormon attachment to home. Perhaps the problem is not apathy toward place but an interest in using place as the definitive context in which a fixed and exceptional identity is formed.[11]

For that matter, I don't know that environmentalists, particularly those of the biocentric school of thought, recognize how ineffective it is to insist that our willingness to harm nature comes from a failure to see how profoundly, if not indistinguishably, our identities are intermingled with the landscapes that we inhabit. Emphasizing our equality with all life and with place in a biocentric or life-centered universe (as opposed to an anthropocentric or human-centered one) always seems to imply that human self-interest is a categorical wrong. It is assumed that if we just accepted our radical equality with all things, we would then logically live on behalf of the whole, even though little evidence suggests this is the case. Moreover, doesn't the fact that we are capable of ethics, especially on behalf of all of life, make us stand apart in the universe? In short, biocentric environmentalism finds the claims to human specialness embarrassing even though, like all moral philosophies, it relies on the careful use of human judgment. Because it can confuse the boundaries of the self with the boundaries of the land, biocentric environmentalism ironically ends up sounding a lot like anti-environmental nativism that simply wants everyone else out. Only a worldview that cannot perceive the outline of our own unique human differences will end up imagining that all interventions on behalf of the land's well-being are invasions into our most private and personal spaces.

Marilynne Robinson's quotation points to the central importance of landscape in our human experience, but it insists that the landscape remains at least partially opaque and apart from our ambition to make it ours. It is ours, she says, only to the degree that we query it, to the degree that the land inspires us to give account of a "particular sense of amazement." I would still like to believe that the eco-theologians are right, that our environmental problem is essentially a problem in our definition of humankind's relationship to the physical world. But because of our long human history of using religion as a justification for our narrow, short-term, and tribal interests, we need to listen more carefully to what religion tells us, especially as it differs from the partisan ideologies of our time.

So my environmentalism has evolved to a very counterintuitive point. I believe that we should care about nature precisely because of our human differences, rather than because of our common heritage of star dust. Ecotheology, arguably the most important development in religion over the last forty years, suggests that environmentalism must divorce itself from partisan politics and must instead stem directly from moral principle. An environmentalism that cannot draw any distinctions at all between the human condition and the physical world risks allowing the ever expandable human ego yet more terrain to lay claim to as its own. By the same token, if religious belief cannot ground its claims of human exceptionalism in the context of interdependency with all of life, then it too will only provide added incentive for human hubris to act perpetually in our own narrow self-interest. What we need is deference, forbearance, patient long-suffering, and a persistent determination to identify the consequences of our choices for which we must repent. These are old-time values that we can all stand by.

On my office window hangs a work of art by a former student of mine, Sarah Judson Walker. It is a stained glass representation of the restoration of the Provo River in Heber Valley, a restoration that took the better part of the 1990s and millions of dollars of mitigation funds from the Jordanelle Dam to complete. No one would argue that the restoration has been perfect, but few believe that the return to a meandering wetland more populated now with osprey and hawks, a healthier fish population, and a growing mammal population was anything but a triumph of engineering, artistic insight, and a newly steeled moral deference for the creativity of life that has nurtured that valley for millions of years. Like great art, ecological restoration is a memorial to a chastened and humble recognition of the world as a surprising gift. Sarah's stained glass piece imagines an aerial view of the restored curvature of the river's bends, freed from the straitening confines of an ethos of private gain. This is water as shared life, as collective rebirth, not water as a commodity. I keep Sarah's work in plain view because as the sunlight pierces through its artifice it reminds me that we still have a chance to restore a deeper artistic and ecological imagination here at home. Maybe standing on the shores of a restored river delta pouring into Utah Lake wouldn't provide the same romance as standing on the edge of the Grand Canyon, but it might provide something more important: confirmation that our own creativity can partner, rather than compete, with the creativity of nature. This would be no small gift to give in return.

Notes

Portions of this essay were previously published online in the March 2012 issue of Provooremword.org and at patheos.com.

1. George Handley and Lance Larsen, "The Radiant Astonishment of Existence: Two Interviews with Marilynne Robinson, March 20, 2004, and February 9, 2007," *Literature and Belief* 27, no. 2: 113–43.
2. William P. Brown, *The Seven Pillars of Creation: The Bible, Science, and the Ecology of Wonder* (Oxford: Oxford University Press, 2010), 4.

3. For information about the history of Mormon environmental practices, see Thomas Alexander's essay "Stewardship and Enterprise."

   For information about recent commitments by the LDS Church, see http://www.deseretnews.com/article/700027829/Mormon-Church-unveils-solar-powered-meetinghouse.html?pg=all. A recent talk by Elder Marcus Nash of the Quorum of the Seventy addressed Mormon stewardship principles (http://ulaw.tv/videos/religion-faith-and-the-environment---marcus-nash/0_93udac8v). Also see the website produced by the LDS church on conservation and stewardship: http://www.mormonnewsroom.org/article/environmental-stewardship-conservation.

   For other writings on Mormon environmentalism, see Terry Tempest Williams, William B. Smart, and Gibbs M. Smith, eds. *New Genesis: A Mormon Reader on Land and Community*. See also George Handley, Terry Ball, and Steven Peck, eds., *Stewardship and the Creation: LDS Perspectives on the Environment*; Gary Bryner, "Theology and Ecology: Religious Belief and Environmental Stewardship"; Matthew Gowans and Philip Cafaro, "A Latter-day Saint Environmental Ethic"; and George Handley, "The Environmental Ethics of Mormon Belief."
4. I have written at greater length about this question in my essay "The Poetics of the Restoration."
5. There is no official statement from the First Presidency of the LDS Church on earth stewardship. The closest thing we have to an official statement on stewardship can be found at http://www.mormonnewsroom.org/article/environmental-stewardship-conservation, a website that includes a link to the aforementioned talk by Elder Marcus Nash of the Quorum of the Seventy. Although statements about stewardship by church leaders are not infrequent and are cited on the website, Elder Nash's talk represents the first time a General Authority of the church has exclusively focused on the topic in a public forum.
6. Hugh Nibley, "Subduing the Earth."
7. Elder Nash states that "according to LDS doctrine, this earth, as well as the plant and animal life thereon, were provided for the use of man. However, we believe that God has commanded that the earth and all things thereon be utilized responsibly to abundantly sustain the human family. . . . He intends man's dominion to be a righteous dominion, meaning one that is guided, curbed, and enlightened by the doctrine of His gospel—a gospel defined by God's love for us and our love for Him and his works. The unbridled, voracious consumer is not consistent with God's plan of happiness, which calls for humility, gratitude, and mutual respect" (http://www.mormonnewsroom.org/article/elder-nash-stegner-symposium).

8. For my understanding of the ecology of the Provo River and Utah and their history, I am indebted to the work of D. Robert Carter and to conversations that I have had over the years with biologists such as Steve Peck, Mark Belk, and most recently Mike Mills, who helped me understand the complex relationships involved in the June Sucker Recovery Implementation Program. See Carter's *Utah Lake: Legacy*.
9. See the website: http://www.junesuckerrecovery.org/.
10. John M. Broder, "No Endangered Status for the Greater Sage Grouse," *New York Times*, March 5, 2010 (http://www.nytimes.com/2010/03/06/science/earth/06grouse.html?_r=0).
11. This was part of my reason for choosing the title of my book, *Home Waters*. I wanted to suggest a way of belonging that didn't have the territorial or anti-environmental implications conveyed by the idea of a homeland. See *Home Waters: A Year of Recompenses on the Provo River*.

5

Climate Change and the Future of Great Salt Lake

DANIEL BEDFORD

Great Salt Lake stands out clearly on maps of North America, looking for all the world like an inkblot on the western United States. It's a cartographic Rorschach test: talk to people about the lake and you can discover more than just their views on the lake itself. To many who live outside the state, the lake is an icon of Utah, as much a feature as great skiing or the Mormon Tabernacle Choir. To many who live in Utah, it's a big, salty, stinky waste of water, useful as a source of industrial raw materials but not much else. To others, it's an extraordinary ecosystem, a scientific enigma to be studied and understood for its own sake as well as to manage the competing uses of the lake and to safeguard a bird habitat of hemispheric importance. Whatever else it may be, though, Great Salt Lake is also a physical manifestation of Utah's past, present, and future water challenges and decisions.

Great Salt Lake is a terminal lake, meaning that it has no water flowing out of it. The amount of water that it contains is therefore almost entirely controlled by the amount of water that goes in from rivers and from precipitation directly onto the lake's surface, minus evaporation. This water balance of inputs minus outputs makes the lake an integrator of Utah's climate: in big snow years with cool springs and summers, the lake level rises; in dry winters with hot summers, the lake level falls. Because Utah's climate is influenced by patterns in the global climate system, the rising and falling of Great Salt Lake indicates global climate events as well as local conditions.

Besides climatic influences, the amount of water flowing into Great Salt Lake is partly a function of society's choices about water, both conscious and unconscious. The more we take out of Great Salt Lake's feeder rivers and don't put back—consumptive use—the less water there is for the lake. Lake level variations are a mirror, reflecting the climate within the lake's drainage basin and beyond and the choices that we have made about whether and how to live in this dry place, adapting ourselves to it and it to us. Consequently, as population grows, as land use changes, as attitudes to water and to Great Salt Lake change (or don't change), and (critically) as climate changes, these dynamics of the societal and physical environment will show up in lake level fluctuations. This chapter is devoted to investigating the links between Great Salt Lake and its societal and physical environment, as they pertain to water. First, it looks at the lake itself in detail, then it examines the lake's connection to regional and global climate and the prospects for climate change in coming decades. In the second half of the chapter, I consider the societal setting of Great Salt Lake and how choices about water being made within the context of a changing climate might further affect the lake.

Great Salt Lake is a remnant of pluvial Lake Bonneville, a very large freshwater lake that began to form around 25,000 years ago in response to climatic changes driven by cyclical variations in earth's orbit around the sun and reached its peak around 16,000 years ago.[1] Although it fluctuated up and down somewhat, leaving distinctive lake shorelines at different levels along the mountains of the Wasatch Front, at its peak Lake Bonneville covered slightly less than 20,000 square miles or about a quarter of the area of present-day Utah. A catastrophic drainage event, a shift in earth's orbit around the sun, and a drying and warming climate made Lake Bonneville shrink to its present diminutive size. As Bonneville lost water to the drying climate, its salts became ever more concentrated in the water that remained: freshwater Lake Bonneville became the highly saline Great Salt Lake. Today this process continues, as the three main rivers entering the lake—the Bear, the Weber, and the Jordan—bring melted snow and tiny concentrations of dissolved salts from the Wasatch Mountains into the broad, flat depression that Great Salt Lake fills. Unlike the situation in more ordinary lakes (which allow salts to pass through them on their way to the oceans), the salts are left behind when the water evaporates from Great Salt Lake, accumulating over time.

It is this salt that defines Great Salt Lake. Placing a number on just how salty the lake is turns out to be trickier than one might expect, however, because salinity is highly variable over time and from place to place. Terminal lakes can experience very large and rapid fluctuations in volume in response to a run of wet or dry years over time, and Great Salt Lake's volume has varied between about 2.5 cubic miles in 1963 and 9.0 cubic miles in 1988 and 1989 over the historical record of lake measurements dating back to 1847.[2] This amounts to nearly a quadrupling in volume. Such massive fluctuations in water volume dissolve and concentrate the salts at high and low lake levels, respectively.

Salinity also varies from place to place. This is partly natural, as salinity is lower where the rivers empty into the lake; but many spatial salinity variations are due to human activities. In 1959 an earth-fill railroad causeway across the lake was completed, forming a barrier to water movement and effectively cutting the lake into a Northern and Southern Arm. The North Arm has no river inflow, so it has become progressively saltier and saltier, until it is now essentially hypersaline. Its water is poised on the brink of saturation, its salt (at least the sodium chloride part) primed to crystallize out of solution as soon as a nucleus, such as a dead insect, presents itself. In very wet years rainfall directly onto the North Arm of the lake can lower the salinity measurably, as it did in the 1980s. But even then salinity remains significantly higher than in the larger South Arm. At average lake levels the South Arm is around three times saltier than the ocean. Smaller causeways, such as the road linking Antelope Island to the mainland, have had similar though less dramatic effects. Much as a mosaic forms a complete picture from many small multicolored tiles, Great Salt Lake is a mosaic of waters with radically different salinities. Depending on the lake level, salinity has varied from around 20–28 percent in the North Arm, to 7–15 percent in the South Arm, to 3–5 percent in Farmington Bay, to 0–7 percent in Bear River Bay.[3]

Salinity matters because the ecosystem of the lake is hugely dependent on it. At average lake levels not much can live in the water of the South Arm besides halophilic microbes, algae, and brine fly larvae and brine shrimp.[4] At the higher salinities found in the North Arm, or when the rest of the lake is affected by prolonged drought, even the brine shrimp start to have trouble. In the fresher water of the river outlets (or elsewhere on the lake during prolonged wet spells) predators and competitors can intrude on the brine shrimp's habitat, thinning their numbers. At average lake levels, though, the brine shrimp have the lake and the algae largely to themselves, leading to population booms that present an irresistible food source for creatures that can reach it—birds.

It is difficult, in fact, to overstate how important Great Salt Lake is for birds, especially the migrants endlessly traveling north and south across the Americas. These birds use Great Salt Lake as a fueling station, resting and loading up on calories for the journey ahead of them. During spring migration, for example, Utah Department of Wildlife Resources bird counts estimate that the populations of some species visiting the lake constitute significant percentages of the North American or even global totals.[5] Quirks in the mosaic of Great Salt Lake salinity add to its appeal for avian visitors: American White Pelicans take up residence each year on Gunnison Island, because it's safe, surrounded by a protective moat of hypersaline North Arm water, and within a day's round-trip flight of the fish living in the relatively fresh water of Bear River Bay, some thirty miles to the east. A survey of the total American White Pelican population over the period 1998–2001 estimated the Gunnison Island colony as the second largest in the world.[6]

Birds aren't the only ones to make use of the lake's food resources. Brine shrimp cysts (eggs) are harvested commercially and shipped around the world to aquaculture operations, where they are hatched to provide live food for larger shrimp that will eventually end up in supermarkets. What began in 1952 as a small-scale business aimed at tropical fish enthusiasts has grown into a multimillion-dollar industry.[7] Because it depends on the maintenance of a long-term supply of brine shrimp cysts, the industry has been a leading supporter of ecosystem research and sustainable harvesting regulations.

Other industries depend on the lake too. Dotted around the shoreline are several commercial salt extraction facilities, operated by companies such as Morton Salt and Great Salt Lake Minerals. They pump lake water into closed-off ponds then harvest the different salts as sunshine evaporates the water. The only U.S. manufacturer of magnesium metal from raw materials, U.S. Magnesium (formerly Magnesium Corporation of America or MagCorp), also operates in this way, using solar concentration of magnesium chloride dissolved in Great Salt Lake water. This is serious, heavy, and sometimes polluting industry: although the company has taken steps to clean up its operations, especially its airborne pollutants, the U.S. Magnesium facility on the western shore of Great Salt Lake was added to the list of highly polluted Superfund sites by the U.S. Environmental Protection Agency in 2009.[8]

Like the ecosystem itself, these industries are strongly affected by the amount of water in the lake, which affects salinity and lake level. Salinity in turn affects both industries: brine shrimping because salinity influences the ecosystem on which cyst harvesting depends, and mineral extraction because more solar evaporation is needed to extract the salts the less salty the lake water is. Lake level directly affects mineral extraction because many facilities are of necessity located close to the water's edge, so rising water can destroy dikes and flood evaporation ponds. Great Salt Lake Minerals facilities suffered a dike breach in 1984 as the lake rose toward its record-high level in 1988, despite spending over $16 million to raise the dikes.[9]

The amount of water in Great Salt Lake therefore matters a lot to both natural and human systems. It can also profoundly affect something as seemingly immutable as the basic physical geography of Utah. The lake is shallow, with generally gently sloping sides, so small changes in the volume of water can produce very large changes in surface area. Terry Tempest Williams puts it clearly in her classic book *Refuge*:

> I recall an experiment from school: we filled a cup with water—the surface area of the contents was only a few square inches. Then we poured the same amount of water into a large, shallow dinner plate—it covered nearly a square foot. Most lakes in the world are like cups of water. Great Salt Lake, with its average depth measuring only thirteen feet, is like the dinner plate.[10]

At their most extreme, the very wet conditions of 25,000 years ago gave rise to Lake Bonneville. In recent history the lake has been as small as about 940 square miles at its historic low level of 4,191 feet above sea level in 1963 and grown as large as about 2,300 square miles at the historic high level of 4,212 feet above sea level in 1988 and 1989.[11]

Great Salt Lake, then, is the sum of its parts, water and salt. Its physical, ecological, and economic characteristics vary hugely depending on how much water is available to dilute the salt. This brings us to the critical question of this chapter (and in many ways of this book): what determines the amount of water available? For Great Salt Lake the water balance equation (slightly simplified) is river inflow plus precipitation minus evaporation—but what controls the terms of this equation? This is a complex area with many unanswered or partially answered questions. For the purposes of imposing some order on this unruly thicket of complexity, the main influences on Great Salt Lake's water balance are natural and human, with some of these factors influencing each other as well as the lake.

The natural influences begin with the lake itself. Although popular perception holds that lake-effect storms are responsible for Utah's famous powder snow, recent research suggests that the lake's role in the quality and quantity of snow falling in its own drainage basin is fairly small. Huge, freshwater Lake Bonneville may have been partly self-sustaining, driving its own lake-effect storms onto the mountains to the east and then recapturing the moisture when rivers brought it back.[12] But Great Salt Lake, much smaller and much saltier, appears to be only a relatively minor factor in driving snowfall, compared with big storms coming off the Pacific Ocean.[13] It's unclear how different the precipitation regimes might be at extreme high and low lake levels, and some research suggests more lake-effect snow from fresh water compared with salty water;[14] but the impact of the lake itself on precipitation appears to be small for many lake conditions.[15] Lake-effect precipitation, it seems, contributes only slightly to Great Salt Lake's own water availability.

A lake feature that does seem to play a role, however, is bathymetry: the shape of the bowl that contains the lake. The lake is shallow, and its sides for the most part slope gently. This tends to translate small changes in volume into large changes in area: when a small amount of extra water is added, the gentle slopes cause the water to spread out, adding a lot of area but not much extra depth. The situation is complicated because lake bathymetry is not uniform: the sides of the bowl that contains the lake slope very gently near the bottom but curve upward more steeply near the top—and this varies from place to place around the lake too. No two parts of the bowl have exactly the same curve from bottom to top. This means that the precise nature of the translation of volume change into area change varies depending on the lake level. This matters because of its implications for evaporation. If the lake were vertically sided, the surface area would remain constant no matter how much water it contained; so the evaporation losses would also remain constant, assuming no change in climate. At the opposite extreme, very gentle slopes make for large increases in surface area for small increases in volume, which greatly increases the evaporation losses relative to the volume of water added. Gentle slopes therefore tend to work against large, sustained changes in volume and surface area, because a small change in volume leads to a large change in evaporation. The sides of the basin that holds Great Salt Lake slope more steeply in some places, more gently in others, which means that the lake tends toward greater stability at certain levels. Statistical analysis of lake-level history, combined with analysis of lake bathymetry, suggests "preferred modes" for lake level at roughly 4,197, 4,201, and 4,210 feet above sea level.[16] These appear to be levels to which the lake tends to settle more easily, and from which it tends to move out of with more difficulty. In this sense Great Salt Lake can be at least a partial master of its own destiny.

Emphasis on the partial: lake bathymetry and its influence on evaporation are plainly not the whole explanation for the lake's water balance. Water input matters hugely: a recent attempt to quantify lake sensitivity to its various water balance terms indicates that river inflow is by far the most important component—in other words, how much rain and snow falls in the drainage basin, running off into rivers and thence into the lake.[17] Wet years and dry years are part of the noise of short-term weather variations, but they, too, seem to have preferred modes. This is where our scale of analysis has to zoom out: instead of thinking only about the lake or only about its drainage basin, we have to think bigger and consider the scale of the global climate system. Ed Lorenz, father of chaos theory as it applies to weather, famously wondered if a butterfly flapping its wings in Brazil could trigger a tornado in Texas.[18] This may be something of an exaggeration for Great Salt Lake's water balance, but the general concept applies: small perturbations in the atmosphere and oceans, generated in faraway places, can affect local conditions. Recent research suggests that wind and ocean current activity in the central equatorial Pacific, together affecting sea surface temperatures, can strongly influence the amount of moisture reaching Great Salt Lake's drainage basin.[19] This is not quite the same as the fairly well known phenomenon of El Niño, a periodic warm ocean current affecting the central and eastern equatorial Pacific. El Niño clearly can and does affect weather in many parts of the world, including southern Utah, but its effects on northern Utah appear to be fairly weak.

The stronger influence for Great Salt Lake seems to come as the equatorial Pacific switches between El Niño and its cold-current sister, La Niña.[20]

All of these influences apply under the current climate. The existence and disappearance of Lake Bonneville, though, tells us that this region of the world has seen radically different climate regimes in the past, both wetter and drier. Some research points to decades-long "megadroughts" at certain periods in the climate history of the southwestern United States.[21] Climate can and does change over time, and as a huge body of research accumulated over decades clearly indicates, human activities are now one of the driving forces of climate change.[22] As we add more and more greenhouse gases to the atmosphere (especially carbon dioxide from burning fossil fuels), we are strengthening its ability to retain and reradiate heat. The level of atmospheric carbon dioxide has climbed from about 280 parts per million (ppm) in 1780, before the Industrial Revolution, to about 316 ppm in 1959, the first complete year of direct instrumental measurements, to almost 394 ppm in 2012.[23] Measurements of the isotopes of carbon accumulating in the atmosphere point conclusively to fossil fuels as the main source.[24] As this has happened, less and less heat radiated from the earth's surface has escaped into space and more and more has been captured, warming the lower atmosphere and earth's surface. This is the phenomenon known as global warming.

Those who are skeptical about the idea that humans are changing earth's climate often rely on misleading arguments, so it's worth being clear on some definitions. Global warming refers to an increase in the earth's average temperature over time, a trend that has been measured with thermometers since around 1860. The Intergovernmental Panel on Climate Change (IPCC) now refers to this trend as "unequivocal": the world has warmed up by about 0.8 degrees Celsius (1.5 degrees Fahrenheit) since thermometer records began, with much of the warming happening since 1970.[25] In summarizing a large body of peer-reviewed scientific literature on the issue, the IPCC has also stated that it is very likely (90 percent chance) that most of the warming of the twentieth century is due to human emissions of greenhouse gases.[26] This global warming trend, however, is just that—a global average. As with any average, it includes changes in temperature that go down as well as up. Some locations are not warming very much or are even getting cooler, but these are outweighed by the many more places that are warming. Although skeptics might point to the rare place where temperatures have decreased, cherry-picking the temperature records of individual places says next to nothing about the global patterns.

This is because the climatic effects of adding greenhouse gases to the atmosphere are far more complex than simply raising temperatures. Adding greenhouse gases effectively adds energy to the climate system, so in addition to a global average warming it is reasonable to expect more heat waves, more intense evaporation and precipitation, and more intense forms of certain kinds of storms, such as hurricanes. We can also expect changes in patterns of atmospheric circulation, which could bring unusually cold, warm, wet, or dry air to specific places, sometimes resulting in counterintuitive weather. For example, some research suggests that melting Arctic sea ice—an intuitive consequence of a warmer world—could contribute to colder, snowier winters in Europe, a decidedly counterintuitive result of global warming.[27] The connection between the surface temperature of the equatorial Pacific Ocean and the water level of Great Salt Lake is another example of how atmospheric circulation ties together distant places in far from obvious ways. For this reason, among others, warming at the global scale results in more complex climate changes at the regional and local scales.

As the earth warms, what might happen to the climate of northern Utah and how might these changes affect Great Salt Lake? Projections of the future state of the climate system are made by using very sophisticated computer models, running on the most powerful computers available in government research institutes all over the world, from Canada to Australia, from Japan to Germany, from the United Kingdom to the United States. The models are imperfect, partly because of limitations in computer power and partly because there are some aspects of climate that remain poorly understood or difficult to represent in the models, such as clouds. To help balance out these imperfections, multiple runs of each model are made, tweaking the variables slightly each time, to give a range of possibilities within the bounds of the physically plausible. Multiple runs of multiple models from multiple different research institutes then give a sense of where the climate might be heading in response to increased greenhouse gas concentrations and how reliable any given aspect of a possible future climate might be: if lots of models and model runs agree quite closely on any given outcome, we can have greater confidence in that outcome than if the models tend to disagree.

Uncertainties in the simulation of the climate itself—whether globally or in those impacting the Great Salt Lake drainage basin—are just one part of the challenge, however. Even more uncertain is where human society might be heading, in terms of energy consumption and its implications for fossil fuel use. Will we discover a cheap, efficient new way of making solar panels, for example, thereby cutting our greenhouse gas emissions? Conversely, will we discover a cheap new way of extracting oil from tar sands, boosting our carbon emissions? What political and economic decisions will we make, individually and collectively, in coming decades, and what impact will they have on the atmosphere?

These questions are almost impossible to answer, so researchers working on projections of future climate don't even try. Instead they work from a set of scenarios, outlining the growth and decay of emissions according to various different futures, broadly optimistic or pessimistic about technological innovation and the globalization that would allow for rapid transfer of technology from rich to poor countries. Because of the near-impossibility of forecasting these kinds of futures, each scenario is considered equally likely. Scenarios used to be known by a series of slightly arcane codenames, such as A1B or A2, but recently the scenario development process has changed to emphasize representative concentration pathways (RCPs). A1B is an oft-cited middle-of-the-road scenario, roughly analogous to RCP 4.5. This change is recent enough that references to both systems of scenario development and naming can be found in the literature.

With this in mind, projections of future climate in Great Salt Lake drainage basin show strong agreement that the region will get warmer in coming decades, with the amount of warming depending on the scenario. There's no shortage of scientific journal articles and reports that reach this conclusion for the western United States as a whole,[28] but I wanted to see for myself what the model statistics looked like. A recent partnership between Lawrence Livermore National Laboratory, the U.S. Bureau of Reclamation, and Santa Clara University has made publicly available a vast storehouse of model results, statistically "downscaled" to try to give a clearer picture of regional as opposed to global possibilities.[29] I downloaded the output from fifteen different models, each one run anywhere from one to six times (with slight tweaks each time), for a total of thirty-nine different estimates of the climate of the future, for the A1B middle-of-the-road scenario and the more pessimistic A2 scenario. The A1B middle-of-the-road scenario suggests summertime warming in Great Salt Lake basin of around 4.5 degrees C (8.1 degrees F) for the end of the century compared with the present. More pessimistic scenarios show slightly more warming—around 5.5 degrees C (9.9 degrees F) for the A2 scenario. Recently published results for multiple runs of one model, known as the CCSM 4, suggest warming for the western United States as a whole of up to 6 degrees C (10.8 degrees F) for the most pessimistic representative concentration pathway, RCP 8.5.[30] While model uncertainties mean that the various different models and model runs disagree with each other about exactly how much warming will take place for any given scenario, the overall level of disagreement is relatively small. Based on this, the scientific community as a whole is reasonably confident that the western United States in general could warm substantially by the end of the century.[31] It is worth pointing out that the global trajectory for carbon dioxide emissions is currently closer to the more pessimistic scenarios than to the more optimistic ones.[32]

In contrast, projections of future precipitation are much more uncertain. While the higher temperatures will mean more precipitation falling as rain than as snow (a shift that has already been observed in the western United States according to several recent studies),[33] there is a lot of disagreement among models and model runs over how much precipitation might fall in the future, indicating the greater complexity of the processes involved. Overall, the A1B model runs suggest wetter winters and drier summers for Great Salt Lake basin, but the level of disagreement from model to model and run to run is high.

Even if the shift in climate is such that northern Utah receives more precipitation overall, however, this is unlikely to translate into more water for Great Salt Lake. Once again, physical environment and societal factors must be considered. A warmer physical environment means more evaporation and greater water demand from plants. So even if the region gets wetter from a climate standpoint, the end result could easily be less water in the rivers, streams, and reservoirs. A1B projections of hydrology for Great Salt Lake basin suggest overall reductions in surface runoff of 10–20 percent, running as high as almost 30 percent in some places. These numbers are derived in part from the precipitation projections, so again they vary considerably from model to model, with some suggesting wetter futures and others suggesting drier ones. But the overall weight of the estimates leans toward a warmer, drier future for the Great Salt Lake drainage basin.

This may well have consequences for the lake itself. Recall that the lake's water balance, simply put, depends on inputs from rivers and precipitation and outputs from evaporation. Less water in the rivers, and more evaporation, makes for a water balance that is shifted toward the negative, making for a smaller Great Salt Lake. A recent thorough study of how sensitive the lake is to changes in the various terms of the water balance found that the lake level could drop by about 2.2 feet, for a 25 percent reduction in river inflow, and by about 1.1 feet, for a 4 degree C increase in temperature.[34] Keeping in mind the caveats about model projections, it seems at least plausible that the lake could be somewhere around three feet lower by the end of the century than it is today, due to the direct physical effects of climate change alone.

These direct physical effects, of course, are not the only ones likely to impact Great Salt Lake under a changed climate. The amount of water reaching the lake is not only a function of the climate; it is also a function of how much water is taken out of the rivers upstream and not returned. It is partly a function of how much water we (as residents of the drainage basin) choose to allow the lake to have. Societal impacts on the water supply to Great Salt Lake might be small compared with the impacts of natural climate variability or human-induced climate change, but they are not negligible. A long-standing estimate of the effects of consumptive water use places the lake level five feet lower than it would be without such use.[35] Related estimates place consumptive use in Utah at roughly half the total water withdrawals for both agricultural and municipal and industrial uses.[36] For the Bear River, the single largest source of water for Great Salt Lake, this equates to almost 14 percent of the total available water in the river basin, if the statewide average ratio of consumptive to nonconsumptive use applies here.[37] Human impacts on the water inflows to Great Salt Lake therefore are currently large enough to take seriously.

As with the effects of climate change, some of which are already apparent, these human impacts look set to grow, perhaps significantly. As discussed elsewhere in this book, the population of the region is projected roughly to double by 2050, with corresponding increases in municipal water demand unless serious water conservation efforts are undertaken—and even then total water demand is very likely to be higher in future decades than today. In some parts of the drainage basin, notably the Weber River basin, it is anticipated that declines in agricultural use will more than offset the increased urban water demand,[38] but this is by no means true everywhere. In the Bear River basin state water managers have noted that a growth in urban population has eaten into the total acreage of dry farmland but not irrigated farmland, so the growth in urban water demand has come on top of, rather than instead of, the water demand for irrigated agriculture.[39] Several municipalities in the Bear River basin are already at or near the limits of their existing water supplies and delivery infrastructure, implying that water diversions will need to increase in the near future.[40] On balance, then, the net impacts of a growing population and shifting land use appear likely to drive an increase in water demand in coming decades. Increased temperatures, driving increased evaporation and therefore increased water demand for irrigation, cooling electrical generating systems, and other uses, may well amplify these population-based changes.[41]

These population changes represent a challenge to Great Salt Lake's water access for two reasons. The first is that the lake does not have a right to water under the standard doctrine for water allocation in the western United States. The second is the basic geographic fact that the lake is the lowest point in the drainage basin: all other uses occur upstream.

Water law in the western United States was established when water was seen in utilitarian terms, for irrigating farmland or extracting minerals, and its central aspects still reflect the attitudes of the nineteenth century. Broadly speaking, western water law has two components: one, prior appropriation must be considered;[42] two, water must be put to beneficial use (which means taking water out of a stream for some purpose) or the right is forfeited—use it or lose it, as the saying goes. These two components conspire against water conservation for natural ecosystems, because they encourage extracting water as quickly and fully as possible from its natural settings. Many water managers and agencies know this and make efforts to mitigate the more harmful environmental effects of western water law while staying within it, but the momentum of the body of water law in the western United States works against preservation of the natural environment. Really, it's all about water rights: whoever started diverting water first has the right to keep on doing so, no matter what impacts this may have on more junior rights holders. In this context it is troubling that Great Salt Lake holds no water rights at all.

Even if it did, at this point it would be a junior water right, low in the pecking order and vulnerable in dry years. Bear River Migratory Bird Refuge, for example, which sits just upstream of the lake on the lower reach of the Bear River, holds a junior water right that often goes unmet because of uses by more senior rights holders upstream.[43] And here, of course, is the second challenge for Great Salt Lake: everything is upstream. Water can be diverted before it reaches the lake, and there are numerous examples of the consequences when this happens to excess. In the 1960s the Soviet Union diverted water from the Aral Sea in central Asia and effectively dried up what was once the world's fourth largest lake.[44] In the United States the Los Angeles Department of Water and Power has diverted water from the Sierra Nevada mountains, drying up Owens Lake in the 1920s and endangering Mono Lake in the 1970s.[45] Although there is no real danger of Great Salt Lake drying up completely, a smaller, saltier lake might be a tougher environment for brine shrimp—think of the limited population in the North Arm of the lake—and birds would be crowded more closely together, increasing the risk of disease outbreaks. More exposed lake bed could mean more dust blown into the atmosphere.

For Great Salt Lake, these difficulties are highlighted by the existing structure for lake management. The Utah Division of Forestry, Fire and State Lands has recently completed a careful, thorough, and detailed comprehensive management plan for Great Salt Lake.[46] The plan outlines the critical resources, areas where more research is needed, and the impacts of rising or falling lake levels. But it is intended for management of the lake at any given lake level; it is not intended as a plan to ensure or provide for a certain lake level. Nor could it be, under existing state law. As the examples of the Aral Sea and Owens and Mono Lakes show, failure to manage upstream conditions threatens efforts to manage the lake downstream, however capable those efforts might be.

Ultimately, the question of how much water we are prepared to give Great Salt Lake is a question of how much we care about the lake. Plenty of anecdotal evidence suggests that the answer to that question might be "not much," at least in the aggregate. For many people, it seems, the lake is a waste of water, as a long history of (unrealized) plans to dike off the lake and create a series of huge freshwater reservoirs suggests.[47] In many ways the lake is perceived as a treasure trove of valuable minerals, with its other aspects remaining secondary. Great Salt Lake Minerals, for example, is moving ahead with plans for significant expansion of its evaporation ponds on the western shore of the North Arm. The plans caused an uproar among many lake scientists when they were first proposed. To its credit, GSL Minerals has scaled back the original proposal to take account of environmental impacts.[48] It took pressure from concerned citizens and scientists to bring about this change, however, which is indicative of a broader mind-set that the lake is useful only for its economic value, as Rob Dubuc notes in chapter 7 herein.

Fortunately, we have alternatives. Instead of seeing Great Salt Lake as a unique resource to be exploited, people are increasingly aware that Great Salt Lake is a unique resource to be conserved. Great Salt Lake Advisory Council, which advises the governor and state agencies on lake-related matters, has recently advocated an integrated water resources management model, in which upstream diversions could be accounted for.[49] As discussed elsewhere in this book, Utahns remain amazingly inefficient water users compared with the rest of the United States, including other dry states. This might be partly explained by Utah's water prices, which are among the lowest in the West.[50] If, as basic economics suggests, water users would cut their consumption in response to higher prices, then Utah water use—and water availability for Great Salt Lake—might be at least partially under our own control; it need not be a prisoner of climate change. By one estimate, Utah currently exports about 5 percent of its total water footprint as "virtual water," embedded in crops and livestock sent out of state.[51] This is a small amount compared with other states but not insignificant compared with the water budget of Great Salt Lake. It indicates the potential for Utah possibly to become a net importer of "virtual" water in the future, freeing up the real stuff for environmental conservation. And the lake always retains the capacity to surprise. When it was near its historic low level in 1963, newspaper articles agonizing over the lake's demise were widespread. By the 1980s, when the lake level had shot up to its historic high, emergency measures to prevent flooding were being rushed into place, including construction of a set of giant pumps on the western shore. The combination of climate change and population growth in northern Utah does not bring good news for the water balance of Great Salt Lake—but it remains a highly unpredictable beast. Although the threats to the lake are real, and the consequences of ignoring them potentially serious, it may be unwise to write its epitaph just yet.

Notes

Comments from Hal Crimmel, Lynn de Freitas, and Carla Trentelman greatly improved the quality of the final manuscript. Any remaining errors of fact or omission remain the sole responsibility of the author.

1. D. R. Currey, G. Atwood, and D. R. Mabey, *Major Levels of Great Salt Lake and Lake Bonneville* (Utah Geological and Mineral Survey Map 73, 1984).
2. Calculated from data in B. L. Loving, K. M. Waddell, and C. W. Miller, *Water and Salt Balance of Great Salt Lake, Utah, and Simulation of Water and Salt Movement through the Causeway, 1987–98*, U.S. Geological Survey Water-Resources Investigations Report 00-4221 (Washington, D.C.: U.S. Geological Survey, 2000).
3. E. Trimmer and K. Kappe, "History of State Ownership, Resource Development, and Management of Great Salt Lake," in *Great Salt Lake: An Overview of Change*, ed. J. W. Gwynn, 523–40.
4. D. Stephens, "Salinity-Induced Changes in the Aquatic Ecosystem of Great Salt Lake, Utah," in *Modern and Ancient Lake Systems: New Problems and Perspectives*, ed. J. K. Pitman and A. R. Carroll, Publication No. 26 (Salt Lake City: Utah Geological Association), 1–8.
5. T. W. Aldrich and D. S. Paul, "Avian Ecology of Great Salt Lake," in *Great Salt Lake*, ed. Gwynn, 343–74.
6. D. T. King and D. W. Anderson, "Recent Population Status of the American White Pelican: A Continental Perspective."
7. D. Kuehn, "The Brine Shrimp Industry in Utah," in *Great Salt Lake*, ed. Gwynn, 259–64.
8. U.S. Environmental Protection Agency, US Magnesium, 2011, http://www.epa.gov/region8/superfund/ut/usmagnesium/; for more details on U.S Magnesium's impact on Great Salt Lake and a broader discussion of pollution's impacts on the lake, see chapter 7 by Rob Dubuc herein.
9. A. E. Isaacson, F. C. Hachman, and R. T. Robson, "The Economics of Great Salt Lake," in *Great Salt Lake*, ed. Gwynn, 187–200.
10. Terry Tempest Williams, *Refuge: An Unnatural History of Family and Place*, 6.
11. Calculated from Loving et al., *Water and Salt Balance of Great Salt Lake.*
12. S. W. Hostetler, F. Giorgi, G. T. Bates, and P. J. Bartlein, "Lake-Atmosphere Feedbacks Associated with Paleolakes Bonneville and Lahontan," *Science* 263, no. 5147 (1994): 665–68.
13. T. I. Alcott, W. J. Steenburgh, and N. F. Laird, "Great Salt Lake–Effect Precipitation: Observed Frequency, Characteristics, and Associated Environmental

Factors"; K. N. Yeager, W. J. Steenburgh, and T. I. Alcott, "Contributions of Lake-Effect Periods to the Cool-Season Hydroclimate of the Great Salt Lake Basin."

14. D. J. Onton and W. J. Steenburgh, "Diagnostic and Sensitivity Studies of the 7 December 1998 Great Salt Lake–Effect Snowstorm."
15. Alcott et al., "Great Salt Lake–Effect Precipitation."
16. I. N. Mohammed and D. G. Tarboton, "On the Interaction between Bathymetry and Climate in the System Dynamics and Preferred Levels of Great Salt Lake."
17. I. N. Mohammed and D. G. Tarboton, "An Examination of the Sensitivity of Great Salt Lake to Changes in Inputs."
18. E. N. Lorenz, *The Essence of Chaos.*
19. S-Y Wang, R. R. Gillies, and T. Reichler, "Multidecadal Drought Cycles in the Great Basin Recorded by Great Salt Lake: Modulation from a Transition-Phase Teleconnection."
20. Ibid.
21. For example, see C. A. Woodhouse, D. M. Meko, G. M. MacDonald, D. W. Stahle, and E. R. Cook, "A 1,200-Year Perspective of 21st Century Drought in Southwestern North America."
22. S. Solomon, D. Qin, M. Manning, Z. Chen, M. Marquis, K. B. Averyt, M. Tignor, and H. L. Miller, eds., *Climate Change 2007: The Physical Science Basis.*
23. NOAA ESRL, "Trends in Atmospheric Carbon Dioxide," http://www.esrl.noaa.gov/gmd/ccgg/trends/.
24. Solomon et al., *Climate Change 2007.*
25. Ibid., 5.
26. Ibid.
27. V. Petoukhov and V. A. Semenov, "A Link between Reduced Barents–Kara Sea Ice and Cold Winter Extremes over Northern Continents."
28. See R. Seager, M. Ting, I. Held, Y. Kushnir, J. Lu, G. Vecchi, H.-P. Huang, N. Harnik, A. Leetmaa, N.-C. Lau, C. Li, J. Velez, and N. Naik, "Model Projections of an Imminent Transition to a More Arid Climate in Southwestern North America"; T. Karl, J. M. Melillo, and T. C. Peterson, eds., *Global Climate Change Impacts in the United States.*
29. S. Gangopadhyay and T. Pruitt, "Hydrologic Projections for the Western United States."
30. S. Peacock, "Projected Twenty-First-Century Changes in Temperature, Precipitation, and Snow Cover over North America in CCSM4."
31. Karl et al., *Global Climate Change.*
32. Peacock, "Projected Twenty-First-Century Changes."
33. R. R. Gillies, S-Y Wang, and M. R. Booth, "Observational and Synoptic Analyses of the Winter Precipitation Regime Change over Utah"; A. F. Hamlet,

P. W. Mote, M. P. Clark, and D. P. Lettenmaier, "Effects of Temperature and Precipitation Variability on Snowpack Trends in the Western United States."

34. Mohammed and Tarboton, "An Examination of the Sensitivity of Great Salt Lake."
35. Great Salt Lake Planning Team, *Great Salt Lake Comprehensive Management Plan Resource Document* (Salt Lake City: Utah Department of Natural Resources, 2000).
36. *Municipal and Industrial Water Use in Utah* (Salt Lake City: Utah Division of Water Resources, 2010). Calculated from table 1, p. 3.
37. Calculated from *Bear River Basin: Planning for the Future* (Salt Lake City: Utah Division of Water Resources, 2004).
38. *Weber River Basin: Planning for the Future* (Salt Lake City: Utah Division of Water Resources, 2009).
39. *Utah State Water Plan: Bear River Basin.*
40. Ibid.
41. Karl et al., *Global Climate Change.*
42. The "first in time, first in right" concept is described in chapter 1 by Eric Ewert herein.
43. *Utah State Water Plan: Bear River Basin.*
44. P. P. Micklin, "Desiccation of the Aral Sea: A Water Management Disaster in the Soviet Union." For a comparison with Great Salt Lake, see D. P. Bedford, "Great Salt Lake: America's Aral Sea?"
45. For example, see M. Reisner, *Cadillac Desert: The American West and Its Disappearing Water.* For more details on this topic, also see chapter 12 by Jana Richman herein.
46. "Final Great Salt Lake Comprehensive Management Plan and Record of Decision."
47. For example, J. W. Gwynn, "The Waters Surrounding Antelope Island, Great Salt Lake, Utah," in *Great Salt Lake,* ed. Gwynn, 107–19; see also Williams, *Refuge.*
48. For more discussion of Great Salt Lake Minerals and evaporation ponds, see chapter 7 by Rob Dubuc herein.
49. Great Salt Lake Advisory Council, *3rd Annual Report to the Utah State Legislature Natural Resources Appropriations Subcommittee,* February 2013, http://www.gslcouncil.utah.gov/docs/2013/feb/gslac_ar_2013_6.pdf.
50. Utah Rivers Council, *Crossroads Utah: Utah's Climate Future* (Salt Lake City: Utah Rivers Council, 2012). For a detailed discussion of water waste and water rates in Utah, see chapter 6 by Zachary Frankel and chapter 13 by Daniel McCool herein.
51. S. T. Mubako and C. L. Lant, "Agricultural Virtual Water Trade and Water Footprint of U.S. States."

6

≈

Chicken Little's New Career

How Utah's Water Development Industry Sows False Fears and Misinformation

ZACHARY FRANKEL

A MASSIVE QUANTITY of water is being proposed for diversion in Utah, more in fact than in any other state in the country. An incredible 650,000 acre-feet of water withdrawals—enough for 5 million Los Angeles residents or 2.6 million wasteful Utahns—are being proposed for diversion across this desert state. Upon learning of any of these proposals, most people naturally focus on how the water will be used. Yet rarely does this approach provide a meaningful perspective, because these projects are not being constructed for the water.

Contrary to popular belief, Utah is not running out of water. Although such headlines sell newspapers and get water projects funded, the notion that Utah is running out of municipal water is a ridiculous myth, perpetuated by and for the benefit of Utah's nearly untouchable water development industry. Utah has never come close to using its full share of Colorado River water, which is but one example of the abundance of available water for municipal needs—the state never built the plumbing to bring this water to the population because the people never needed it. Even the Central Utah Project, for example, delivers less than 10 percent of the water used in the entire nine-county area it serves, with the rest coming from other water suppliers. In fact, city dwellers—98 percent of Utah's population—are using just 8 percent of Utah's total water inside and outside their homes.

Utah is proposing these water diversions because of the dysfunctional nexus between big-money politics and wasteful water economics. No

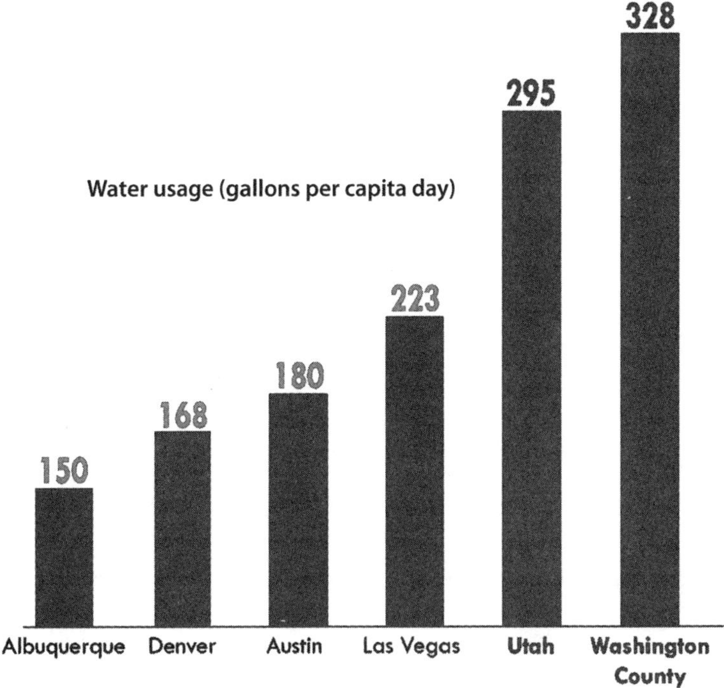

Figure 6.1. Municipal water use in the West. Utah Rivers Council.

single policy exemplifies this dysfunction and demonstrates the real purpose of these water projects more than Utah's widespread practice of water suppliers who collect property taxes.

It's nearly lunchtime at the State Capitol on a late summer morning. In a conference room by the cafeteria, three camera operators place TV cameras on tripods alongside two newspaper photographers. All wait patiently for the Utah Rivers Council press conference to begin. We're pleased that two reporters have come in person to hear the story, because we have something waiting outside the door that's about to dazzle these professionals accustomed to staid events.

These days few news outlets have the money to send reporters into the field, and we feel fortunate to have a strong turnout. Ten years ago a good press conference could bring a dozen reporters from TV, radio, and print outlets. They had real experience working the beat and tended to ask tough questions tinged with skepticism. The final story would be better for it.

Often the net result was better policy decisions—or at least more informed ones.

Today press conferences are mostly about attracting camera operators, who often have no idea what room they're supposed to shoot in, much less knowledge of the story itself. If reporters do show, chances are they've never reported on water before, because the turnover in media outlets is high and the reporting experience is often minimal. The polite explanation for this new era is that "people get their news from the Internet," which results in a less informed populace and budgets that don't allow for veteran reporters to get into the field as often.

I'm wise enough to know that these reporters did not come to hear me speak, so I ask our guest of honor to join us. As soon as he walks in, the room comes alive. Cameras pan, following his every movement, and jockey for the best shot. It's not every day that a man in a six-foot chicken suit covered with yellow costume feathers wearing a tweed sports coat and reading glasses comes to offer his wisdom about Utah water policy at the State Capitol. But this man is Chicken Little: not only has he walked out of a folktale to join us, but he has been studying Utah water politics of late.

Mr. Little steps to the podium, clears his throat, and explains his presence at the Utah Legislature. "When I heard that St. George is America's most wasteful water user, and the solution to this waste is to tell people they are on the verge of running out of water, and need $1 billion to waste more water, I just had to come offer my support," said Mr. Little. "I mean, the brilliance of charging residents the cheapest water rates in the American West to encourage people to waste water while simultaneously scaring them into believing they are running out of water and need to spend $1 billion-plus to divert the Colorado River—well, I knew I had something to learn from these good old boys."

It's clear from the confused laughter that some don't quite understand Mr. Little. But they seem to enjoy hearing his much-needed perspective on state policy. "So I took a break from my important work alerting everyone that the sky is falling to see how Utah elected officials could let these water development industry guys propose to raise yet more taxes—in addition to the property taxes they pay for water—for a completely frivolous purpose," he opines. Reporters hang on his every word.

"I must say I haven't been disappointed," the Chicken concludes. "In fact, I've been so inspired by this fear and delusion that I feel like it might

Figure 6.2. Chicken Little at the Utah State Capitol. Utah Rivers Council.

be time for me to make a career change and get into the water development industry."

Mr. Little provides an electric presence in the room, so the crowd is predictably disappointed when I take the podium to provide data to explain why Mr. Little strongly objects to the use of irrational fear in planning Utah's water supply. I unveil a new economic analysis completed by a group of economists referenced in a letter to Utah legislators signed by a dozen academics from the University of Utah, Utah State University, and Dixie College that explains the basic problem with this water project: it costs a lot of money, and it isn't needed.[1]

When the Utah Legislature passed the Lake Powell Pipeline Development Act, it included a requirement that the construction costs be repaid with interest by the recipients of the water. In other words the State of

Utah—the taxpayers of Utah—will loan money for all planning and construction costs to the water suppliers building this expensive project. Then the customers of the government agencies that deliver this water will repay all costs incurred with interest back to the taxpayers.[2]

The economists analyzed the long-term borrowing and spending proposal outlined in the Washington County Water District's Capital Facilities Plan. The analysis includes Excel sheets, pie charts, and long-term revenue predictions to answer the basic question: can this agency repay the $1 billion-plus in construction costs required to build the pipeline? With only around $12 million in annual net revenues and roughly $10 million in current annual debt payments, the agency is in no condition to dig itself into more debt. The project would burden this agency with a whopping $47 million in additional annual debt payments for fifty years, far eclipsing even its gross revenues, much less net revenues.[3] For this small St. George water supplier with big dreams, the Lake Powell Pipeline is a bottomless pit of debt, I tell the assembled reporters and TV crews.

Calculating debt limits by institutional borrowers is similar to the mortgage formula used by the local bank for homeowners. The revenues available to repay a loan every year should be a fraction (typically 30 to 40 percent) of the homeowner's take-home pay or an entity's net annual revenues, in the vernacular of commercial lending. The economists determined that the Washington County Water District is proposing to burden its customers with annual debt payments equivalent to nearly four times its current net income.

Commercial lenders and your local banker would never loan borrowers more than their annual net incomes, much less four times their net income, I remind listeners, because total annual debt payments should be a fraction of net income not a multiplier. In other words, even water supply agencies should make more money than they spend. But then this lender doesn't have to make sound financial decisions because this lender is the Utah Legislature.

I continue by noting that the Utah Legislature is besieged by water development industry lobbyists who scare legislators into believing we are on the verge of shriveling into prunes in the desert sun unless the Lake Powell Pipeline, or virtually any other water project for that matter, is constructed. There are many interests who stand to profit immensely from a billion in new spending and are happy to manufacture a fictitious water crisis for legislators. In fact, the biggest advocate for the Lake Powell Pipeline boondoggle is a legislator who runs a water agency that would receive water from the pipeline. With the prospect of big money on the table, whether the water is needed or whether or not the costs can be repaid seems unimportant to this lobby. And even if "environmentalists" have crunched the numbers, their perspective is irrelevant because this is a water project and therefore a pure good, goes the thinking: discussion over, let's spend money.

The room has grown silent, but I continue, noting that the barrier standing in the way of the pipeline proponents is that Washington County has taxed itself into a corner. The Washington County water supplier gets 41 percent of its total annual revenues not by selling water but by collecting property taxes on the value of local homes. These property taxes subsidize the price of water by allowing the government water supplier to lower the price of water far below what it costs to deliver it to people's faucets.[4] These property taxes are the root cause of Utah's high water use, because of basic market economics.

I show the audience that St. George residents pay $1.11 per thousand gallons of water at the peak of summer water use in July in comparison to Seattle residents, who pay $15.74 for this same volume of water even though Seattle is one of America's wettest cities, where water is more abundant. Las Vegas residents are charged four times this rate, a whopping $4.58 per thousand gallons of water. Las Vegas's per capita water use seems Spartan compared to that of St. George. In the water "need" documents for the proposed Lake Powell Pipeline, the State of Utah noted that Washington County's water use was 328 gallons per person per day, more than twice the national average and nearly 100 gallons more than Las Vegas residents' use, according to the Southern Nevada Water Authority.

These cheap rates allow people living in an arid region to rank among America's biggest water users per capita and therefore as water wasters. Consider that Las Vegas employs water cops who issue citations with fines if they catch a homeowner putting water onto sidewalks or into the gutter. In contrast, though—and as the audience well knows—on any summer day it's hard to find a neighborhood in Utah's cities that isn't watering pavement, with several gutters funneling water through the street. This is not water use, it's water waste. The recipients of water from the proposed Lake Powell Pipeline water (Washington and Kane County residents) already are the undisputed winners in the world of water waste, wasting more per capita than people anywhere else in the United States, thanks to the property tax for water. Water suppliers and the Utah Division of Water Resources then claim that we are running out of water, without implementing substantive conservation measures.

At first glance it seems as if the Washington County Water District could simply raise water rates to repay Utah taxpayers. But basic market economics dictate that the more expensively a commodity is priced, the less of it consumers will purchase. The economists' analysis is that in order to repay the construction costs, revenues would have to be increased a stunning 370 percent, meaning that much higher water rates would be required in Washington County to repay the construction costs from the loan. Assuming that voters and their elected officials could somehow magically grow enthusiastic about a steep increase in their water rates, this still wouldn't solve the repayment problem.

If water rates are raised dramatically to increase revenue, I suggest, people will reduce their water consumption dramatically, which in turn will reduce the total revenue available to repay such a massive loan. Reducing water use also creates a surplus of water and defers the need for additional water, which eliminates the original rationale for the Lake Powell Pipeline. Doing more with less has led to thousands of successful conservation programs across the globe, including projects in water, energy, natural gas, oil, and electricity. This concept, however, is still novel to many Utah water suppliers. Adding to this problem, the fact that seven out of ten gallons of water used in Washington County are for agriculture, according to the Utah Division of Water Resources, makes it clear that the St. George area has an abundance of water for municipal use. If Washington County should urbanize as much as local leaders claim it will, these agricultural lands will be converted to urban lands, providing an abundance of water available for new residential and commercial development.

As I stop providing facts and figures, I realize that all these data may be putting the audience to sleep. It's time for the closing statement: "The best alternative to this expensive water project," I say in my most dramatic voice, "is simply to phase out property taxes for water." Alas, because I am not wearing a six-foot yellow chicken suit adorned with red feet, reporters have started to tune out. Camera operators position themselves to compose shots featuring a man in a chicken suit next to a talking head at the Capitol. Despite the facts, people still seem to think that the sky is falling. As if I needed more evidence that my data were upstaged by the giant chicken and the crisis mentality that it represents, when I open the floor to questions one of the reporters asks, "What makes you think that St. George isn't running out of water?"

It's a wintry morning outside Professor Lindsey Christensen Nesbitt's crowded Global Change Ecology class at the University of Utah. I begin every lecture by asking the same question, "How many people think we're running out of water?" On this morning over one hundred hands shoot up. "Okay. Now how many people have actually walked over to the tap, turned it on, and no water came out?" I ask. All but three hands come down. "And how many of you had your water run out *not* because of a burst pipe or forgetting to pay the water bill?" I ask.

All hands come down. "That's interesting," I note with feigned surprise. "Okay, how many people know someone else who has run out of water and not because of a plumbing or payment issue?" I ask. No hands are raised. "Well, that's strange, isn't it?"

"Now I'm going to make a pie chart of Utah's total annual water use by dividing all our water use into just two categories: agriculture and everything else, also known as municipal-industrial use. Who can tell me how much water agriculture uses each year as a percentage of the total, compared to all other water uses by humans?" To help confuse them, I add: "Let me give you a hint: 98 percent of Utah's population falls into the municipal-industrial sector for the purposes of categorizing water use. Another 2 percent of the population falls into the agricultural sector, which consists of farmers and ranchers. What percent of our total water is used by this tiny fraction of the population?" An empty pie chart on the screen awaits an answer.[5]

After some coaxing, students start guessing: "80 percent," "70 percent," "40 percent." The correct answer comes rarely: often only one person among a pool of hundreds across several lectures can provide it. "About 85 percent of the water used in Utah is used by agriculture," I say.

With a few mouse clicks pushing data on the screen, I walk the students through the one piece of information that I wish every Utahn knew about water in this very dry state: 98 percent of Utah's population is using just 2–3 percent of Utah's total water inside their homes on an annual basis.

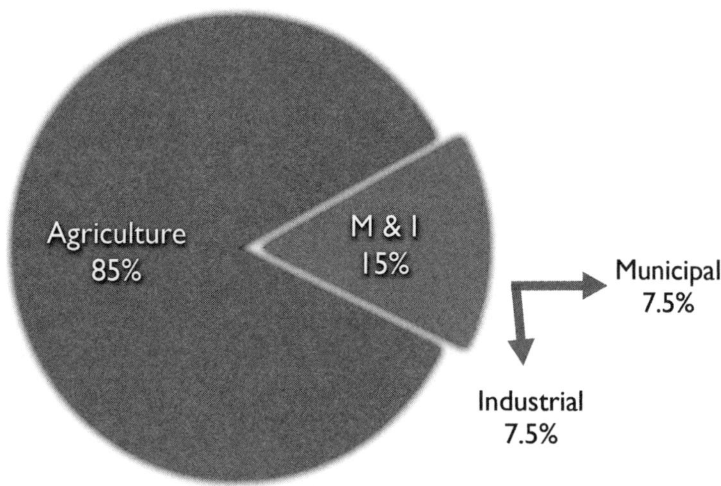

Figure 6.3. Annual water use in Utah. Utah Rivers Council.

This is a water need. And even with 98 percent of Utah's population using 6–8 percent of Utah's total water outside their homes,[6] primarily during the summer months to water grass, which is a nonessential water use, it's clear that we are not running out of water. "How many people still think that we're running out of water?" I ask. This time only a few hands are raised. "Before my explanation, nearly 100 percent thought that we were running out of water. I wonder why so many thought that?" I ask.

The students' embrace of fear as fact illustrates how successful water development industry marketing has been in making such sales pitches to the public. In a *Salt Lake Tribune* guest editorial, the general manager of Washington County Water Conservancy District makes his case for spending a billion dollars in taxpayer funds on the proposed Lake Powell Pipeline. The project has become increasingly unpopular because of its massive price tag and because water lobbyists seek to fund the project by taking money away from the funding for popular government services such as public and higher education, law enforcement, corrections, and health services. The first line of his editorial perfectly articulates the go-to strategy to get any water project built in Utah: "Disappointment for home-owners is turning the tap and no water gushes forth."[7]

This highly paid general manager fails to note that he has no idea how much water could be saved if he phased out his agency's property tax col-

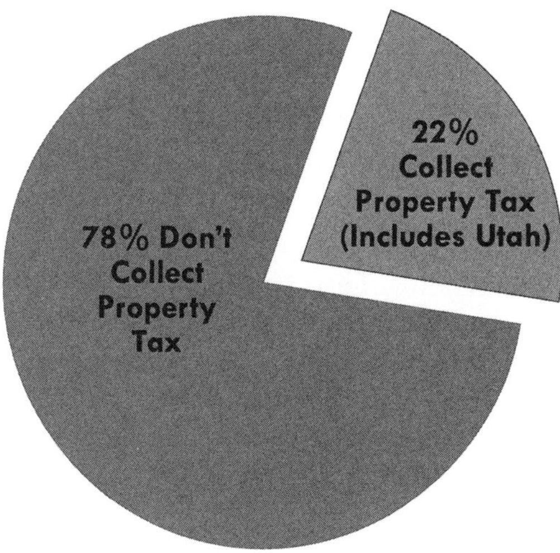

Figure 6.4. Western water districts collecting property taxes. Some 78 percent of western water suppliers in the United States do not collect property taxes, based on a survey among the fifty-eight major urban water delivery agencies. Utah Rivers Council.

lections.[8] Phasing out property taxes would lower per person water use dramatically and provide large quantities of water, but this sensible alternative to spending $1 billion-plus is not mentioned.[9]

The manager also fails to mention that Utah's property tax subsidies for water are unique in the American West. Roughly 80 percent of the nearly sixty water suppliers surveyed in the metropolitan areas across the eleven western states do not collect property taxes.[10] Yet every Utah water supplier will assert that taxes are essential to operations, a statement at odds with the free-market economics that Utahns claim to honor. Other utilities required by Utah homeowners such as electricity, telephone service, and natural gas are provided by private companies who receive no property taxes. Yet somehow water can't be sold to customers without collecting property taxes? Privatization is hardly a panacea; but when governments use fear as an advocacy tool to implement policy, bad things happen.

In Utah's water-development industry, fear of running out of water has been saturating legislative committee dialogues, media stories, and opinion pieces like this one for decades. The industry has been so successful at promoting fear that most residents have no idea that Utah is America's most wasteful water user.

This fear campaign is led by the Utah Division of Water Resources, a little-known state agency without much oversight by state legislators. The division subscribes to the notion that something has to die when you conserve water through a water conservation program, because all water in Utah is fully appropriated.[11] This claim implies that there is no water waste in Utah, which again is nonsense given how hard it is to find a U.S. city using more water (per capita) than Utah's cities.

Utahns not only use nearly twice the national per capita average of municipal water but use more water per person than residents of Denver, Phoenix, Las Vegas, Albuquerque, Seattle, Portland, Los Angeles, or San Diego—basically every city in the western United States.[12] In general Utah water consumers receive little incentive to conserve water, due to low water rates and active discouraging of conservation practice. For instance, in a 2012 article comparing Utah's cheap water rates to those of other western cities, Utah Division of Water Resources director Dennis Strong opined that "I would hate to see us get into a situation where only people with lots of money can have grass and trees."[13]

This is a surprising statement, because past planning studies from the division's other work, for instance, indicate that the Salt Lake Valley alone could reduce its water use by 20 percent without removing one blade of grass.[14] Many Salt Lake Valley residents overwater their lawns, and water is priced below its true market value.

Yet the division seeks to inject a measure of fear into its future water use projections for the Salt Lake Valley, writing: "Currently developed supplies are not sufficient to meet project growth."[15] The document reiterates this statement more than twenty times in the first sixty pages, a pace that is maintained throughout the document. At no point does the 115-page document discuss how water use would decline if property taxes were phased out, thus providing water for new uses.

If Utah truly was running out of water, employing basic market economics to save water would be the first step to reducing waste. Phasing out property taxes around the state that bundle the cost of water would allow water rates to rise naturally, with water use declining as a result; this would be a conservative, market-driven principle for water that would seem in keeping with Utah's core values as a state. Many economists and fiscal conservatives have pointed this out to the Utah Legislature, to no avail. The irony is that Utah's conservative elected officials bend over backward every election season in claiming that they will lower Utahns' taxes if elected but refuse to embrace a market-driven approach to water.

It isn't surprising that the water districts are opposed to reducing their revenue streams, even though it would eliminate the need to build expensive capital projects like the Lake Powell Pipeline, the proposed Bear River development project, and others. But these projects aren't about delivering water. They are about building the empire of local water suppliers and generating big building contracts.

It's a smoggy morning outside an Environmental Science class taught by Professors Christine Clay and Brent Olson at Westminster College. It is early in the morning, so I'm speaking extra loudly: "I call it Chicken Little because it's what the Division of Water Resources uses to incite fear of a nonexistent water crisis. Perhaps the most shameful thing about this exercise is that it violates freshman math—sorry, not freshman college, freshman high school math."

On the screen is a graph with the *x* axis representing time in decades and the *y* axis representing total water used by Utah's cities. A line inclining upward magically predicts where Utah's municipal water use outgrows the available supply of municipal water. It crosses the supply line at the place where we "run out of water"—the place that Utah water buffaloes work hard for you to have nightmares about.

These water demand figures are not magic; there is no complicated spreadsheet of calculus, no team of engineers programming a supercomputer to print out lines on a graph. The slope of the line depicting future water use is based on some guy multiplying two numbers together: population growth times per capita water use.

"As you can see," I show the students, "the graph indicates that their guess for future water use was based on the per capita use of 295 gallons per capita day from the year 2000. So let's look at their data set."

The next slide is a bar graph from the Division of Water Resources showing nine years of municipal water use in Utah. I ask the students if anyone can tell me what's wrong with using the year 2000 data point as the basis for this projection.

The entire room gets it, but a very bright young woman in the front excitedly raises her hand first: "It's the single highest year of data, not the average of water use for the entire period."

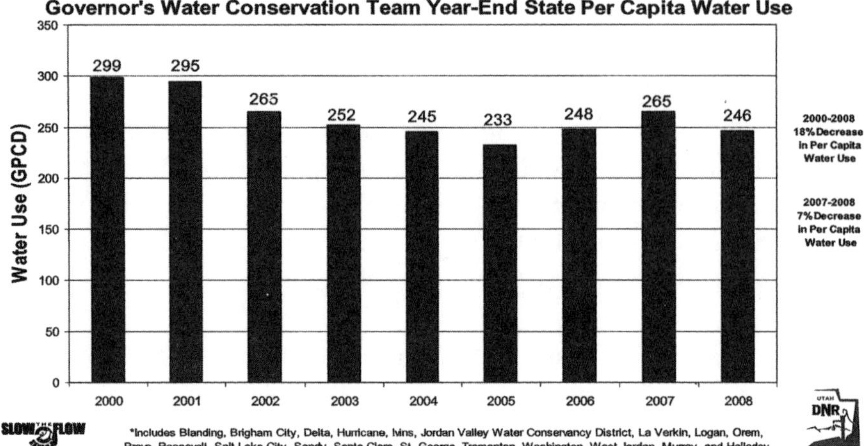

Figure 6.5. Utah per capita water use.

"Exactly," I say. "They picked the single highest water-using year from the entire nine-year period intentionally to inflate future water needs. Either that or the planners are unfamiliar with the concept of averaging." I make quotation marks in the air with my fingers. "If they use an average of these nine years, they would reduce water use 18 percent—just by using a calculator. It's as if they intentionally want to misinform us into believing that we are running out of water. What's interesting about that 18 percent figure is that it's the same amount that the Division of Water Resources claims we have reduced our water use by in the last thirteen years."

A few days later, on yet another smoggy winter day, I am running late for a panel discussion at Weber State University. On the panel is an individual from the Weber Basin Water District, which seeks to spend $2 billion on proposed Bear River development.

This project, currently on the drawing board, would be the final straw in more ways than one for Great Salt Lake, which is the last great wetland ecosystem in the American West. The lake contains roughly 500,000 acres of wetlands, which support habitat for more than 250 bird species traveling from every country in North, South, and Central America and as far west as Siberia. As indicated elsewhere in this collection, more than 5 million individual birds, including several species that gather in larger numbers than anywhere else on the planet, depend upon the Great Salt Lake's wetlands for their survival.

The proposed Bear River diversion would divert nearly 20 percent of the river's annual flow to the lawns of Weber, Davis, and Salt Lake County. The Bear River provides between 58 and 70 percent of the Great Salt Lake's surface water inflows each year, so the diversion is widely expected to lower the water level of Great Salt Lake, thereby drying up shoreline wetlands around its perimeter, perhaps by tens of thousands of acres or more.[16]

It is telling that the Weber Basin Water District delivers three times as much agricultural water as municipal water. Part of this water delivery is for secondary water, untreated water delivered through canals that are relics from when this region was primarily agricultural. It is equally telling that this secondary water is delivered to homeowners to flood-irrigate grass. Virtually no one knows how much water is used because there are no meters. A planning document from the Division of Water Resources indicates that water is frequently overused in this unmetered system by 100 percent.[17]

Nonetheless, the representative of the Weber Basin Water District uses the tried and true fear tactic that we all know and love: "I would hate to see the day when we walk over to the tap and turn it on and nothing comes out."

Although many legislators consistently stress the importance of Utah's agriculture, in truth many of these leaders view farmlands as undeveloped future suburban real estate waiting for the right developer with "a vision." That's why legislators and the governor consistently refuse to put money into Utah's open space funding program, the LeRay McAllister Critical Land Conservation Fund, even in surplus budget years.

Every day in Utah nearly thirty acres of Utah farmland are converted to subdivisions and strip malls. This rate of farmland loss has been occurring for the last twenty-five years, according to the American Farmland Trust.[18] We don't irrigate our carpets, parking lots, driveways, strip malls, or sidewalks (I hope), so these new urban users demand much less water per square foot than the thousands of acres of irrigated farmland that used to be there. That's why urban growth on farmland creates a net surplus of water.

But when developers convert farmland, the water delivered for urban use is almost always treated, high-quality water, in separate pressurized culinary systems from different diversion points. More often than not the canals are left where they were built when the land was growing alfalfa, fruit, or wheat. Canal companies continue to operate the canals, which is why many suburban neighborhoods and shopping malls across Utah's suburban Wasatch Front still have canals carrying surplus water. The Salt Lake Valley alone has some forty or fifty canal companies still operating.

The Division of Water Resources Jordan River Plan acknowledges the abundance of water held in speculation in the Salt Lake Valley. It's estimated that well over 120,000 acre-feet of water, enough for a city of half a million Utahns or 1 million residents of Los Angeles, flow through various canals in the Salt Lake Valley each year. But only 5,000–10,000 acres of farmland remain in Salt Lake County, according to one source,[19] meaning that speculators are holding onto this water.

It is a warm October morning in the village of Mt. Pleasant. This rural town in central Utah lies at the western toe of the Wasatch Plateau, an 11,000 foot high, 6 mile wide uplift straddling the Great Basin and the Colorado River Basin. Mountain headwaters of alpine ponderosa forests and aspen meadows give birth to the scenic Price River, an important tributary to the Green River, which itself supplies about 40 percent of Colorado River's inflow as measured at Lake Powell Reservoir.

The Price River is one of my favorite rivers in Utah. It starts off as an innocent clear trout stream draining the snowy peaks to become Gooseberry Creek. As it flows northward it joins the blue ribbon trout fishery appropriately called Fish Creek and descends through summer foraging meadows for Rocky Mountain elk and steep limestone canyons of subalpine firs before it enters Scofield Reservoir, one of Utah's most popular sport fisheries. Downstream the Price River becomes a rowdy teenager as it enters a gorgeous steep ravine, held captive between Highway 6 and the railroad. It's a wild Class IV and V whitewater run that has forced many a cocky kayaker to take an adrenaline-fueled swim for safety in between holes, ledges, and steep, violent chutes.

The Price River matures when it reaches the valley floor and the coal mining towns of Helper, Price, and Wellington, providing the water for both agriculture and municipal use. This amazing river then becomes an elder of liquid poetry as it rambles forty miles through the sparse and stoic Book Cliffs in a 1,000-foot-deep roadless gorge before giving itself to the Green River. The priceless canyons of the Green provide rare spawning habitat for the Colorado River watershed's top predator—the Colorado pikeminnow—a miraculous ancient fish that can grow six feet in length and thrive in flows from 50 to 500,000 cfs, with remarkable eyes that allow it to see through the Green's muddy flows.

We are here in Mt. Pleasant looking for the headquarters of the Sanpete County Water District, a government agency living entirely off the collection of property taxes. For several decades this entity has been proposing to dam the headwaters of the Price River with a 17,000 acre-foot reservoir and divert its precious flows into the Great Basin, ostensibly to provide a third crop of alfalfa hay for Mt. Pleasant farmers.[20] At least that's the stated purpose of the project and the one that most people are willing to accept. The Bureau of Reclamation's last Environmental Impact Statement (EIS) on this bad idea acknowledges that only one gallon of every two gallons diverted for agriculture to the area's farms would actually be used for growing crops, a 50 percent conveyance loss. This loss could be eliminated with simple modernization efforts for existing water delivery systems that cost pennies on the dollar compared to the $30–70 million construction cost of the proposed dam.

The proposed trans-basin diversion would send Price River water into the Great Basin and diminish the water supply for Carbon County, which has earned enmity toward the project from those coal towns and resulted in several lawsuits against the Sanpete Water District. The project would destroy much of the blue ribbon trout fishery, diminish the water quality in both Scofield Reservoir and the Price River, and threaten current agricultural operations in Carbon County. The diversion would also dry up flows each year in the lower Price River gorge and the spawning habitat of the endangered Colorado pikeminnow, among other species.

I am pondering my anger toward all this senseless destruction, sitting in the passenger side of a pickup truck owned by our research director Gordon Rowe, who is inside city hall trying to find an address for the water district. The government entity has no board of directors listed, no meeting notices, no meeting minutes, and no physical address beyond a Post Office box in this sleepy town. We're here to find out when the final EIS will hit the streets and just look them in the eye to understand who we're dealing with.

Gordon has been in there for a long time and I'm starting to wonder if he mentioned the Utah Rivers Council, an organization that might not be regarded kindly by these Mt. Pleasant folk. City Hall is also the sheriff's office. As I stare at the patrol car next to us I consider entering the building to extract Gordon. After what seems like an eternity Gordon returns to the truck and does what I had previously thought impossible: he lowers my respect for the worst of Utah's water suppliers.

"Okay," Gordon says, "you're not going to believe this." His face is glowing with a strange mixture of shock and humor. "The water district doesn't really exist."

"What?" I ask.

"They don't have an office or any headquarters and they don't have any staff." He smiles at me, waiting for me to get the punch line.

"What do you mean? Don't they have a reservoir or a treatment plant or a billing center?" I ask.

"They don't have anything," says Gordon. "They have no facilities, no plant, no trucks—nothing. They have never delivered a drop of water before. The guy in there told me that his uncle is on the board and they meet in his living room from time to time. He offered me their phone number if I wanted to talk with him. He was really nice."

"But these guys have been collecting property taxes from residents in this county for twenty-plus years—what do you mean they don't have an office or staff? Are you kidding me?" I ask, feeling somewhat foolish that even I have been naïve about the nonsense of Utah's water bureaucracy.

"I'm totally serious. That's why we can't find anything here."

I suddenly feel as if we have stolen some priceless piece of information that has been under guard for decades. Essentially we have. "We'd better get out of here," I say.

As we drive out of the tiny town of Mt. Pleasant up the steep slopes of the Wasatch Plateau questions come tumbling out from both of us. Where did all the money go? Who's looking over their shoulder? Who's making decisions about how the money is spent? Why would these people want to pay property taxes to an agency that doesn't exist?

We crest the gorgeous plateau while deer scatter on either side of our truck. I imagine Chicken Little coming to that charming little town and renting a small farmhouse by paying in IOUs. He doesn't worry about making a living because he's got property tax revenue flowing into his coffers, without providing one single drop of water to these naïve taxpayers. He's got his work cut out for him. The whole county's residents are so convinced that a water shortage exists that they are willing to put property taxes into a nonexistent agency with a nonexistent staff, a nonexistent office, nonexistent trucks, all to address a nonexistent water crisis. But it's even worse than that.

Property taxes paid by homeowners and businesses are collected by each county's assessor, who sends the money on to the local water supplier. That is, unless the water supplier has not sent the required financial reports to the Utah state auditor. These financial documents are meant to provide a level of financial oversight and include an audited financial statement conducted by a third party. If the state auditor does not receive the necessary financial documents, he or she directs the local county assessor to withhold these property taxes until they are received.

The state auditor ended up withholding the property tax collections of the Sanpete Water District for three years in a row because this "agency" never sent the necessary financial documentation. In the fourth year, 2009, it filed one year of financial records, allowing it to receive money both for 2009 and for all three years of previous property tax collections at once. Then in 2010 this mysterious group of people filed no financial records and had their property taxes withheld again.[21]

As we drive along the gorgeous Price River we contemplate the incredible irony of the property tax paradox. The good people of Sanpete County pay property taxes for four out of five years during the worst economic downturn in eighty years for no services whatsoever to an agency that doesn't exist but parks the money in the county bank so that it can collect dust. These taxpayers live in one of the most politically conservative areas in the country and elect officials because they claim to be against taxes, hate wasteful government, and consistently criticize environmentalists for trying to stop what they consider to be an essential service. All because they think they are running out of water. The chicken, with his fear tactics, is so clever.

Notes

1. Gail Blattenberger and Gabriel Lozada, "Economic Analysis of Washington County Repayment for the Lake Powell Pipeline" (2012; paper in author's possession).
2. Lake Powell Pipeline Development Act, 2006, http://www.water.utah.gov/lakepowellpipeline/WordDoc's/LPPDevelopementAct.pdf.
3. Ibid.

4. Gail Blattenberger, "The Price of Water in Salt Lake County" (1998; paper in author's possession).
5. "Municipal and Industrial Water Use in Utah," Utah Division of Water Resources, December 2010.
6. "Population Data Based on Economic Activity," State of Utah, Governor's Office of Planning and Budget.
7. Ronald W. Thompson, "Thompson: St. George Deserves a Pipeline," *Salt Lake Tribune*, July 14, 2012, http://www.sltrib.com/sltrib/opinion/54479455-82/pipeline-utah-lake-colorado.html.csp.
8. This general manager is the highest paid government official in all of Washington County as of fiscal year 2012, according to the public record. See "Public Employee Salaries," Washington County Water Conservancy District, http://www.utahsright.com/salaries.php?city=water_wash.
9. "Mirage in the Desert," Utah Rivers Council, Salt Lake City, Utah, 2002.
10. Ibid.
11. This axiom has been referred to as "Strong's Law" in public presentations, after Dennis Strong, director of the Utah Division of Water Resources. Strong has claimed that he was misquoted. See Bob Nelson, "Water Conservation Coalition Calls for Audit of Utah Division of Water Resources," Kuer.org, June 27, 2013, http://kuer.org/post/water-conservation-coalition-calls-audit-utah-division-water-resources.
12. Data based on phone surveys and figures provided on municipal websites: for example, for Albuquerque, http://www.cabq.gov/progress/environmental-protection-enhancement/dc-34/indicator-34-2.
13. Amy Joi O'Donoghue, "The Fight for Water: Can the Mighty Mississippi Save the West?" *Deseret News*, May 13, 2012, http://www.deseretnews.com/article/865555735/The-fight-for-water-Can-the-mighty-Mississippi-save-the-West.html?pg=all.
14. "Municipal and Industrial Water Use in Utah," State of Utah, Department of Natural Resources, Division of Water Resources, December 29, 2010, 16.
15. "Executive Summary," *Jordan River Basin Planning for the Future*," State of Utah, Department of Natural Resources, Division of Water Resources, 2010, xiii.
16. For more detail on this issue, see chapter 3 by Craig Denton, chapter 5 by Daniel Bedford, and chapter 7 by Rob Dubuc herein.
17. "Municipal and Industrial Water Use in Utah," 2010, 20.
18. "Utah Statistics," American Farmland Trust, http://www.farmlandinfo.org/statistics/Utah.
19. Jordan River Basin Water Plan, Utah Division of Water Resources, 1998.
20. Mark Havnes, "Thirsty Farmers in Sanpete County May Get Dam Soon," *Salt Lake Tribune*, May 23, 2010, http://www.sltrib.com/news/ci_15147031.
21. Utah state auditor, personal communication, 2011; Utah state auditor, personal letter, 2010.

7

Time to Rethink Policy

Ideas for Improving the Health of Great Salt Lake

ROB DUBUC

IF YOU TOOK a poll on the streets of Salt Lake City and asked people what they thought of Great Salt Lake, you would likely be surprised by the answers. People from out of town are likely to describe the lake as inspiring, a very special place. The reaction of local people is likely to be quite different: they don't think about the lake; they haven't been out there since they were children; it's buggy; it's stinky...it's simply there.

From the perspective of the average Utahn, Great Salt Lake is a misunderstood resource. It's not often perceived as the treasure it really is—something to be preserved and protected. It's some abstract thing shimmering off on the horizon at sunset or possibly a place to take an out-of-town guest—if you can figure out how to get out there. In short, the average Utahn knows very little about the lake, except perhaps that it causes lake-effect snow and that it stinks if the wind is just right.

But those who really know the lake scientifically or viscerally (because their profession draws them to it or because it is where they play) have a passion about the lake that is decidedly at odds with the way the average resident views it. Those who know the lake seem to be in on a well-kept secret. The difference in perception is astoundingly vast. One of the biggest challenges for those working to protect the lake is to change those limited perceptions and to share the lake's secrets so that others come to share a passion for it.

To speak with those who have experienced the lake for the first time—and I mean immersed themselves in the experience, not just stood on the shoreline—is to realize that you're speaking with a convert. There's no going back. People who have spent a day on Antelope Island trails or have braved the wetlands to spend time among the birds or have sailed across the lake or ridden an airboat out on Willard Spur discover a world so intense as to be surreal, as if they have been transported to some magical, mysterious location. "*This* is Great Salt Lake?" they ask. How could they be so close to home and so far from home at the same time? There really is no place quite like the lake.

The problem, of course, is that because the lake is so misunderstood it's also underappreciated and thus mistreated. This is especially true of Utah's politicians, who view the lake in terms of dollars rather than seeing it for the precious resource that it is. "There's money in that there water" seems to be the perception. Lots of money, as it turns out. The lake contributes $1.32 billion in economic output to the state's economy annually, of which $1.13 billion comes from mineral extraction.[1]

But Great Salt Lake is a terminal lake: all excess water from within the Great Salt Lake watershed—whether fresh from the mountains or polluted by urban storm runoff—flows into the lake and remains there.[2] In other words, the lake has no outlet. Because most of the water flowing into the lake comes from snow melt, how much will flow into the lake in any particular year is anyone's guess, given variability in precipitation.[3] And because the amount of snow falling within the watershed varies annually—with recent swings tending toward extremes—so too do the depth and size of the lake. From high to low the cumulative variation in lake level spans 21 feet, from a low of 4,191 feet above sea level in 1963 (24 feet maximum depth) to a high of 4,212 feet above sea level in 1987 (45 feet maximum depth). The "average" lake level is normally cited as 4,200 feet (33 feet maximum depth),[4] although it's likely that the average lake level is trending downward in a time of fluctuating climate and precipitation and with predictions of lower precipitation in the West.

The issue of fluctuating water levels in a lake is not uncommon in times of drought, but these fluctuations are a normal part of Great Salt Lake's ecosystem because it's a terminal lake. While the fluctuations are normal, the long-term downward trend is not. This is important: as the level of the lake declines, so does its value as a resource for both recreation and mineral extraction. Even without considering the effects of a warmer and drier climate, increasing development pressure along the Wasatch Front will continue to subtract from water flowing into the lake. Threats to divert other inflows, such as the Bear River,[5] do not promise a bright future for the lake.

The issue of whether the state has an obligation to prevent water levels in the lake from dropping below a certain minimum in order to protect the public trust values of the lake is complicated and beyond of the scope of this chapter. But because the lake is owned by the citizens of Utah as a sovereign land,[6] and because the state is the trustee responsible for protecting the values of its sovereign lands, many feel that the state should be required to maintain a "conservation pool" of water within the lake. Where this water will come from in a time of increased demand and decreased inflows is at the center of the controversy.

This chapter outlines current issues facing Great Salt Lake and offers an overview of both the characteristics of the lake and the threats to its long-term health and survival. Some readers may be unfamiliar with the pressures on the health of the lake's ecosystem, such as nutrient loading in Great Salt Lake wetlands as a result of growth along the Wasatch Front, the billion-dollar business of mineral extraction, the impact of various industrial discharges, the problems created by diking and causeways that resulted in the formation of distinct water bodies within the lake, and the implications of the lack of water rights for the future of the lake's health.

The vast majority of wetlands in Utah—about 85 percent—are part of the Great Salt Lake Ecosystem.[7] Almost all of those wetlands are found along the eastern shoreline of the lake. As development along the Wasatch Front has crept closer and closer, more of these wetlands have been threatened both directly by development and by a product of growth—nutrients. Because of the fluctuating nature of water levels within the lake, a number of waterfowl management areas, both public and private, consist of wetlands that allow waterfowl managers to control water levels within those areas.

In some portions of the lake, especially Farmington Bay and the wetlands along the southern portion of the shoreline fed by the Jordan River, the amount of nutrient load contained within the wetlands has led to a situation where the algae mat fed by these nutrients has choked out all other aquatic life. It has also helped fuel the explosive growth of an invasive form of phragmites (reed) within these wetlands.

The source of this nutrient load is varied, but much of the blame can be placed on Publicly Owned Treatment Works (POTW), otherwise known as sewage treatment plants, that service the population centers along the Wasatch Front. In the Willard Spur, along the northern shore of the lake, the slower pace of growth has precluded the need for POTWs. But those conditions are changing as more and more homes are built north of Salt Lake City. These two areas along the lake's eastern shore, the Willard Spur and the Jordan River delta, illustrate two very different approaches to this problem—one trying to ensure that the impacts associated with nutrients will not occur and one trying to clean up the mess after the fact.

In the summer of 2010 the Utah Division of Water Quality (DWQ) proposed to permit the new $28-million Willard/Perry sewer treatment plant, which was slated to discharge into the Willard Spur of Great Salt Lake.[8] Once it served as a delta of the Bear River where it enters Great Salt Lake. Today the source of water into the Willard Spur consists of the portion of the Bear River that overflows from the Bear River National Wildlife Refuge. As a practical matter, this means that the refuge uses whatever water it needs from the Bear to fulfill its mission. Anything left over then flows into the lake. In a typical year little or nothing flows into the spur from the refuge from early June to late October because of agricultural use of the Bear upstream. This lack of water combined with summer evaporation results in a situation where a good portion of the spur is reduced to mudflats during the summer months. In about half of the years it is completely isolated from the rest of the lake.

Even though the spur experiences this artificial seasonal drying cycle, it is considered to be one of the most productive portions of Great Salt Lake because of the relatively unpolluted and fresh water. The proposed discharge from the sewer treatment plant threatened to have a significant impact on the wetlands of the Willard Spur, including the unimpounded segment of the Bear River National Wildlife Refuge. Because of these threats a challenge was brought to the permit for the treatment plant, claiming that DWQ had not adequately considered the effects of nutrient loading on the Willard Spur.

At the heart of the concern was the damage that's been done to wetlands along the eastern shore of Great Salt Lake to the south of Willard Spur, resulting from nutrient loads that come at least in part from sewer treatment plants. Prior to construction of the treatment plant, nutrient discharges to the spur had been limited to agricultural and stormwater runoff, with minimal adverse impact. In its consideration of the challenge, DWQ modeled the possible impacts from nutrient loads emanating from the treatment plant and concluded that the discharge could cause significant degradation of the spur within five years.

Because DWQ concluded that the Willard Spur was, in the agency's own words, a "reference quality" wetland worth protecting, the agency rescinded its permit approval and asked the Utah Water Quality Board to fund both additional treatment measures at the treatment plant and a study to examine possible long-term impacts to the spur as a result of the discharge. As a result of this request, the board agreed to set aside approximately $1.5 million to fund a three-year study that will lead to a recommendation for a protective designation for the spur. That study will be concluded in 2014.

But the Willard Spur is not the only Great Salt Lake wetland threatened by nutrient loading.[9] Because Great Salt Lake is a terminal lake, most if not all of the pollutants discharged by the POTWs throughout the Great Salt Lake watershed end up in the lake. Compounded by agricultural and urban runoff, these discharges represent a significant source of nutrient load to the Great Salt Lake ecosystem and have substantially altered the character of lake wetlands—especially the impounded wetlands in Farmington Bay. DWQ held off submitting the state's 2008 303(d) impaired water body list to EPA for approval until 2011. This was largely because DWQ would have been required to declare the majority of Great Salt Lake wetlands impaired, based on the then existing regulatory provisions related to pH and dissolved oxygen. To avoid that DWQ argued that the numeric criteria that it had been using for the wetlands didn't accurately reflect the health of those waters. It therefore rescinded the offending regulations and implemented new ones that allowed the agency to consider a multitude of factors in determining whether lake wetlands were impaired.

Although the conservation community was reluctant to support a rule change that removed any of the numeric criteria protections that existed for the lake, DWQ made it a point to consult with the conservationists early in its rule-making process and continues to involve the public in its attempt to assess and improve the health of lake wetlands. Because DWQ has been straightforward and open throughout this process, local conservation groups made the decision to work with DWQ on this issue rather than challenge its decision to forego listing the wetlands as impaired. According to the agency, the new regulations will be designed to assess wetland health in a comprehensive fashion using a multimetric analysis. Although DWQ has not committed to a firm deadline for completion of this work, the agency maintains that it will be finished sometime in the next few years. Unfortunately, in the meantime no actions have been taken to reduce the discharge of the nutrients that end up in Farmington Bay and elsewhere. These wetlands continue to be plagued by algae blooms and the loss of submerged aquatic vegetation.

But nutrients affect the various species of plant life differently throughout the Great Salt Lake wetlands. One plant in particular, phragmites, has invaded massive stretches of the lake along its eastern shore.[10] Sometimes referred to as a reed, phragmites has a tall stalk topped with a white, feathery plume; certain types are used in decorative landscaping. The strain of phragmites that threatens the health of the lake wetlands (*P. australis*) is non-native and extremely aggressive. Over the past thirty years the area that *P. australis* covers along the eastern shore has increased from 20 percent to 56 percent. Once introduced, *P. australis* invades quickly and can change marsh hydrology and alter wildlife habitat. All told, roughly 22,000 acres of wetlands along the eastern shore of Great Salt Lake have been infested with phragmites, choking out all other vegetation.

Eradication methods generally consist of treatment with herbicide, followed by burning. While this may kill the targeted phragmites patch, however, these efforts do not restore the native wetland vegetation. Because of the aggressive nature of the plant rhizomes, the root balls that occupy the area of the lake bed while the plant is alive actually remain once the plant has been killed off. In order to restore the wetlands after treatment the area must also be cleared with machinery. Local conservation groups are becoming increasingly alarmed over this threat to the wetlands and feel that the State of Utah is not dedicating sufficient resources to fight the problem—especially given the amount of revenue generated by royalties from both the brine shrimp and mineral industries based on the lake.

One of the biggest sources of royalties is mineral extraction, with the various companies involved in that business generating $1.13 billion in total economic output on an annual basis and creating over 5,300 jobs.[11] Given the amount of minerals in the lake, some concede that a certain amount of mining is appropriate, while others feel that these industries have already taken over huge segments of the lake—enough is enough. When one of the largest mineral extraction companies proposed tripling its footprint on the lake, the question became not so much whether mining has its place but whether we've reached a saturation point beyond which we begin to damage the lake's ecosystem.

Great Salt Lake Minerals Corporation (GSLM) currently has approximately 45,000 acres of mining operations split more or less evenly between Bear River Bay and the North Arm of the Lake. In 2008 GSLM submitted a proposal to triple the size of its operation, expanding its footprint on the lake by 91,000 acres. In order to support this expansion the company would also triple the amount of water that it consumes, threatening to lower levels in the lake significantly.

Not surprisingly, this proposal was met with a wave of opposition from a variety of groups concerned about the impact that the expansion would have on the lake. From the perspective of local environmental groups, the lake has been subjected to death by a thousand cuts. For some, any further industrial development was unacceptable. Other groups expressed concern about the impact on lake levels of the extensive withdrawal of water and what impacts those withdrawals would have on wetlands along the perimeter of the lake. Still others had concerns about impacts on Bear River Bay, within which the company planned an 8,000 acre expansion, and on Gunnison Island in the North Arm.

Since 2008, the company has been working with the U.S. Army Corps of Engineers to review the expansion plan under provisions of the Clean Water Act. To do this, the corps began assembling an Environmental Impact Statement (EIS), to study the environmental impacts of the expansion plan. Due to delays on the part of GSLM in providing material, the corps suspended the EIS process during 2012 but restarted in the spring of 2013 when the company submitted its new, downsized proposal.

Based largely on pushback from the conservation community, GSLM met with conservation groups to see if it could address a number of the environmental impacts. One major concern was the impact that the increased water withdrawal would have on the North Arm. Because the North Arm is isolated from the rest of the lake by the Lucin Cutoff railroad causeway (discussed in detail below), the withdrawal would have had a disproportionate impact on water levels in that part of the lake. In low lake level years, substantial lowering of the lake in the North Arm would allow a land bridge to be formed to Gunnison Island, facilitating predator access to a vital American White Pelican rookery.

Additionally, the sheer size of the proposed expansion caught a number of groups by surprise. The cumulative impact of occupying 80,000 acres within the bed of Great Salt Lake was a matter of concern. A 72,000-acre expansion in the North Arm, for instance, would consume miles of shoreline. While the North Arm is largely devoid of wildlife in an average year because of salinity levels reaching 26 percent, during high water years such as the 1980s the North Arm was an essential haven for birds.[12] This is because the increased amount of water in the North Arm diluted salinity levels to the point that brine shrimp thrived in that location, while at the same time salinity levels in the South Arm were too low for the shrimp population.

Moreover, as explained below, plans are being made to replace the existing railroad causeway culverts, which are currently failing and have been a constant maintenance problem, with a bridge. At least in theory that should achieve a better mixing of lake waters between the North Arm and South Arm. To the extent that this is successful, the mudflats, shoreline, and waters of the North Arm will become increasingly important to wildlife.

Finally, the proposed 8,000-acre expansion in Bear River Bay threatened a fundamental change in the dynamics of one of the most productive portions of the lake. Although the bay dries out due to agricultural water withdrawals throughout the summer, the bay is teeming with life during spring runoff. Flightless molting Canada Geese depend on the open expanses and meandering sheet flow within the bay for survival, and the submerged aquatic vegetation that fills the bay serves as a vital food source for waterfowl. The initial GSLM proposal not only would have occupied a significant and productive portion of the bay but would have channeled the current sheet flow in a way that would have created a fast-moving stream.

After years of controversy, and a series of meetings with interested stakeholders, GSLM proposed an expansion plan early in 2013 that will be much less damaging to the lake's ecosystem than the company's initial proposal. First, by increasing the efficiency of its process and by lining the exterior of its outer dike walls, GSLM announced that it would withdraw its application for the 353,000 acre-foot water right that it had submitted to the Utah state engineer as part of its initial expansion proposal. Second, GSLM was able to obtain a substantial quantity of private land on the west shoreline contiguous to the lake that allowed it to shrink its footprint within the confines of the lake. Third, rather than expand into Bear River Bay, the company is foregoing its lease option on that land and will instead expand its operations around Promontory Point. All told, the footprint for the new expansion proposal of 37,500 acres is roughly half of the initial expansion plan, uses no additional lake water in the evaporation process, and steers clear of Bear River Bay. While the questions of whether the environmental analysis conducted by the corps is sufficient and how the company will mitigate for its expansion remain unanswered, GSLM should be given credit for listening to the concerns of the conservation community and modifying its initial proposal.

GSLM occupies a significant portion of the northern portion of the lake. If you use Google Earth to look down on Great Salt Lake, however, you'll notice that the entire southwest portion of the lake is diked off into a series of evaporation ponds. This area of the lake, formally known as Stansbury Bay, supports a number of extraction industries, including salt producers such as Morton and Cargill as well as U.S. Magnesium, which produces all of the primary magnesium metal in the United States. As Great Salt Lake Minerals looks to expand—taking more salt and water from the lake than it currently does—there is an inherent tension between GSLM and the south-shore industries. Because GSLM extracts its brine from the North Arm, with 26 percent salinity, and the south-shore industries extract their brine from the South Arm, with 12–18 percent salinity, the south-shore industries would understandably prefer to see more North Arm salt exchanged with the South Arm.

Co-located with the south-shore mineral companies, U.S. Magnesium occupies 4,500 acres along the western shore of the lake. As noted, U.S. Magnesium produces all of the primary magnesium metal in the United States and provides 14 percent of the world supply. The company extracts magnesium chloride from the lake and splits the magnesium from the chlorine. After several years of litigation, the EPA formally designated U.S. Magnesium as a Superfund site in August 2011.[13] The site contains high levels of environmental contamination of metals (such as arsenic, chromium, mercury, copper, and zinc); acidic waste water; chlorinated organics; polychlorinated biphenyls (PCBs); dioxins; and polycyclic aromatic hydrocarbons (PAHs). Several local groups will be involved in the oversight of the investigations and feasibility studies that will be conducted at the facility. Friends of Great Salt Lake, a watchdog group committed to protecting the lake, has received a technical assistance grant from the EPA to help fund an expert who will consult with the various groups and participate in the Superfund process on their behalf.

Because Great Salt Lake is a terminal lake, any industrial discharge generated in the watershed will end up in the lake unless treated. This is especially true for the largest copper mine in U.S. history, Kennecott Utah Copper (KUC), whose processing facilities are located directly adjacent to the lake. It's clear that the type of discharges that occur from Kennecott's mining operation would be handled differently if Great Salt Lake were a freshwater body. The lake remains a mystery in many ways because of its nature, and water quality standards remain elusive. As a result discharges are allowed to occur into the lake that contain much higher levels of toxic substances, such as selenium, than would be allowed otherwise. The debate continues on what levels of contaminants should be allowed and what harm is being caused by these discharges. Until additional resources are dedicated to studying the effects of these discharges, that controversy will remain unresolved.

Kennecott Utah Copper is located along the eastern edge of the Oquirrh Mountains in the Salt Lake Valley. KUC is the second largest copper producer in the United States, producing 18–25 percent of the nation's copper.[14] The major issues stemming from Kennecott's operation relative to Great Salt Lake result from the tailings impoundment along the south shore of Gilbert Bay, the discharge into the bay from the tailings impoundment, and contamination of groundwater resources in the Salt Lake Valley. Additionally, concern has been growing that a groundwater plume containing high levels of selenium from KUC legacy contamination of aquifers near south-shore facilities may be discharging into Gilbert Bay.

The current footprint of KUC's two tailings impoundments is 8,900 acres along the south shore of the lake. Each year KUC transports approximately 60 million tons of tailings to the impoundments through a 14-mile concrete slurry pipe, which results in an increased height of the impoundment walls on the order of 8–10 feet every year. Beyond the public safety issues associated with the potential collapse of the impoundments during an earthquake, the major impacts of the impoundments are the displacement of wetlands, direct discharges to Gilbert Bay from the tailings pile, and possible discharges to the lake from groundwater contamination. The company has recently sought a permit to expand its impoundment areas by 1,992 acres, 721 acres of which are considered to be waters of the United States.[15]

The water used in KUC's slurry system comes from a variety of waste streams from the company's mining operations, including copper tailings from the Copperton Concentrator; slag tailings from the slag concentrator at the smelter; power plant ash slurry; smelter process waters; wastewater effluent slurry from the hydrometallurgical plant at the smelter; mine leach water and meteoric contact water partially treated in the tailings pipeline; wastewater effluent concentrate from the reverse osmosis treatment of acid/metals sulfate-contaminated waters; neutralization of acid/mine-contaminated waters; Barneys Canyon mine pit dewatering and heap leach pad drain-down waters; construction, maintenance, and lunchroom trash; and treated effluent from the sewage treatment plant.[16] Despite this mixing of wastes, the permit for the impoundment discharge applies only to technology-based effluent standards for discharges from mines that produce ores from open pit operations. When the safe carrying capacity of the impoundment for water is exceeded, water that is not returned for use into the slurry system or is used for dust suppression is discharged into Gilbert Bay. The primary discharge constituent of concern in that discharge is the toxic pollutant selenium. While the company is discharging at levels below what it is permitted, that limit and how it is enforced remain controversial.

Kennecott's past mining activities in the Salt Lake Valley created significant legacy contamination of surface and groundwater in the western part of the valley. As the result of a Superfund settlement, the company has taken a number of actions, including building a reverse osmosis (RO) plant on its own property and providing funding for the construction and operation of a second RO facility that will provide clean drinking water for area residents. Eventually the waste stream from both of these RO facilities will be discharged into Great Salt Lake, which is a point of contention for some conservation groups.

While direct discharges of selenium from various sources into the lake have been documented, a full one-third of the annual load of selenium into the lake remains unaccounted for. In attempting to account for this loading, several scientists have found increasing evidence that points to a possible groundwater plume, containing significant amounts of selenium, that could be discharging into the southern part of the lake. While the source of this plume remains uncertain, a connection to the Kennecott mine is suspected. While these data are preliminary in nature and may not fully account for the missing amount of selenium loading into the lake, conservation groups are continuing to encourage state and federal agencies to investigate this possibility.

No small part of the challenge of protecting the water quality of the lake is that except for selenium no numeric water quality standards exist for Great Salt Lake. Numeric water quality standards are tools used to ensure that specific pollutants in discharges are limited to levels deemed safe to water bodies. Numeric standards developed for freshwater bodies, however, do not apply to saline water bodies such as Great Salt Lake. In fact, one of the challenges in developing numeric standards for saline water bodies is that each is unique in composition, unlike freshwater bodies. Therefore a standard developed for Mono Lake, for instance, wouldn't apply to Great Salt Lake. Currently the one numeric standard for selenium applies only to Great Salt Lake (and even then only to the Gilbert Bay portion of the lake).[17]

Recently DWQ published a draft strategy for tackling the issue of establishing numeric standards for the lake as a whole.[18] While this is certainly a very positive step and should eventually lead to meaningful numeric standards for the lake, it will take a considerable amount of time and resources before that happens. In the meantime DWQ will continue to apply narrative standards to discharges into the lake. Just as the name implies, narrative standards are descriptive standards, used in permitting decisions and very general in nature. An example is the narrative standard for Bear River Bay: "Protected for infrequent primary and secondary contact recreation, waterfowl, shore birds and other water-oriented wildlife including their necessary food chain."[19]

The application of such a narrative standard is necessarily subjective. For instance, what amount of a specific pollutant can be allowed in a discharge and still protect waterfowl? Although DWQ is making a sincere effort to protect the lake and conservation groups are doing their best to work closely with DWQ in that effort, a natural tension exists. Another factor is out of the control of either of these groups—entities seeking discharge permits. Industry has and always will do only what is required of it to clean up its externalities or, in other words, the pollution which it creates. With numeric criteria that requirement is clear. But with the type of narrative standards applicable to Great Salt Lake, it is uncertain where a particular line should be drawn. There will continue to be tension between those who wish to protect the lake against harm and those who wish to discharge pollutants into it.

In an attempt to address the potential threat that discharges of selenium posed to Great Salt Lake birdlife, in 2004 DWQ convened the Great Salt Lake Water Quality Steering Committee to recommend a site-specific numeric standard for selenium for the open waters of the lake. The result of the effort was a selenium standard for Gilbert Bay only. The standard is based on the concentration of selenium in egg tissue and, according to DWQ, is evaluated based on the geometric mean of eggs sampled during a single nesting season. The standard set by DWQ for that geometric mean is 12.5 mg/kg, which would result in a mortality rate in mallard eggs of 10 percent. The U.S. Fish and Wildlife Service took the position that this standard violated the Migratory Bird Treaty Act and recommended to EPA that it disapprove DWQ's recommended standard. In spite of FWS's concerns, however, in December 2011 the EPA approved Utah's selenium standard.

Regardless of whether the egg tissue standard is acceptable, the standard itself is unenforceable. In order to determine compliance with the standard, a sufficient number of eggs must be collected along the south shore of the Gilbert Bay. Currently no more than 25 percent of identified eggs can be collected for this purpose. DWQ has been unable to collect what it terms a "statistically meaningful" sampling of eggs from that area. Additionally, some of the eggs that have been collected show evidence of deformities that cannot be explained by the concentration of selenium in the tissue, according to DWQ. As a practical matter, until DWQ translates the egg-tissue standard into a numeric water column standard and effluent limits, Utah's selenium standard exists in concept only.

Changes have come about not only from discharges into the lake or the mining of minerals but also because several causeways have been constructed across areas of Great Salt Lake, effectively carving it into separate water bodies. The two main causeways across the lake are the causeway to Antelope Island and the Lucin Cutoff. Although the Antelope Island causeway has more or less made Farmington Bay a freshwater bay, its creation in 1969 has not proven to be controversial for the conservation community. This is perhaps because the causeway has made it possible to access Antelope Island State Park, a true state treasure. Contrasted with this is the Lucin Cutoff, which has effectively rendered the North Arm of the lake a completely separate water body.

The Lucin Cutoff, also known as the Union Pacific Railroad causeway, was originally constructed in 1904 as a trestle across the northern portion of Great Salt Lake. It was designed both to shorten the distance that trains were required to travel around the lake and to help avoid the steep grades of the Promontory Mountains north of the lake that were part of the original transcontinental railroad route.[20] The Golden Spike—the symbolic connecting of the Union and Central Pacific railroads at Promontory Summit in 1869—is due north of the Lucin Cutoff.

While the original version of the Lucin Cutoff allowed for a full interchange of lake waters between the North Arm and South Arm, the trestle construction proved unstable and was replaced in the late 1950s by a rockfill causeway. The replacement causeway was originally designed to allow the exchange of brine between the two arms. Because of settling and the addition of significant materials during the high-water years of the 1980s, however, the body proper of the causeway no longer allows that interchange. The only meaningful interchange has been through the two largely ineffective 15-foot culverts and a 300-foot breach that is often above the level of the lake. Citing structural failure of the culverts, the railroad closed the west culvert permanently in 2012 and was recently granted permission by the Army Corps to temporarily close the east culvert and move forward with construction of a 180-foot bridge. The breach to accommodate the bridge is ostensibly designed to replace the flow from the two culverts, although this claim has not been verified.

Numerous studies have been conducted indicating that salinity levels in the North Arm began to rise shortly after construction of the rockfill causeway. Beyond precipitation, there are no meaningful inflows into the west side of the lake. This lack of inflow, combined with the physical separation of the Lucin Cutoff, has led to a situation where waters of the North

Arm eventually reached saturation (about 26 percent) and typically remain there. In fact, significant amounts of salt have precipitated out of the North Arm onto the floor of what is also known as Gunnison Bay. The exception to this was during the mid-1980s, when lake levels rose well above flood stage, resulting in a significant dilution of the North Arm. For several years the salinity levels in the North Arm were comparable to current South Arm levels. It was the only portion of the lake where brine shrimp were viable (brine shrimp require salinity levels of 12–18 percent).

Some critics have stated that conservation groups won't be happy until the causeway is removed. Like it or not, however, people increasingly recognize that completely removing the causeway under current conditions would probably change the lake ecosystem in unexpected and unhealthy ways. With lake levels close to the historic low, the sudden influx of the tremendous amount of North Arm salt would increase salinity levels in the lake as whole beyond what could be tolerated for species such as brine shrimp. Rather than advocating a complete removal of the causeway, current thinking is that some method of managing exchanges between the arms, such as a flexible breach system, would be preferred.

Perhaps one of the biggest threats to the lake in the future is whether sufficient water will remain in the lake to support its many uses. As is typical under western water law, water that is not used in a beneficial fashion—as that term is defined by state law—is considered wasted and is subject to appropriation by a user who *can* use it beneficially. While it seems obvious that the resources of Great Salt Lake would be of little use without water in the lake, a senior state official recently stated in a public forum that he would be fine if all of the water was taken from the lake if it was needed for other uses. Given that the lake is a $1.3 billion annual industry, such a position is hard to justify. Still, unless something is done, the water in Great Salt Lake has no legal protection.

Given the amount of industrial and residential development occurring along the Wasatch Front, it's not hard to imagine that increases in water use as a result of this development will have a negative impact on the lake. In fact, a case could easily be made that the issue of water quantity rather than water quality is the biggest threat facing the lake going forward. Mono Lake and Owens Lake come to mind as examples of how water exploitation has a significant negative impact on an inland water body in the West.

In its most recent version of the Comprehensive Management Plan for the Lake, the Utah Division of Forestry, Fire and State Lands outlined impacts to the resources at various lake levels.[21] This effort clearly shows that at very low levels the resources of Great Salt Lake are significantly impacted. While the literature currently shows the "average" level of the lake to be 4,200 feet above sea level, some have suggested that this level should be lowered to reflect the reality of increased water demands for development along the Wasatch Front. Additionally, several climate change studies have suggested that water throughout the West will become scarcer over time as a result of the warming climate, resulting in lower snow pack and spring runoff. With the addition of the proposal to damn the Bear River to provide culinary water to the Wasatch Front, it's almost a certainty that the lake will become stressed due to this combination of water withdrawal and lower spring runoff.

The solution is something that is not likely to be warmly received by existing or potential water users. A minimum lake level needs to be protected by law. Among the various groups involved in this discussion, this concept is known as a "conservation pool." The idea is to set the level of the conservation pool at an elevation slightly higher than a level where many of the lake resources are significantly impacted. For example, at an elevation of approximately 4,193 feet, a land bridge is formed to Gunnison Island in the North Arm, which would allow predators such as coyotes to access the American White Pelican rookery on that island in a way that would destroy that resource. In this case the ideal conservation pool level would be an elevation that prevented that from happening. As another example, once the lake reaches an elevation of approximately 4,195 feet, the harbors on the south shore and on Antelope Island become inaccessible to sailboats of a certain size. Again, if the intent is to protect navigation, it is important to set the conservation pool at a level that would prevent this from occurring.

Even if officials accept the validity of that concept—which is far from a certainty—two main issues must be resolved. First, at what level should the conservation pool be set? Second, how will the level of the pool be maintained? While the first question poses some challenges, setting that level becomes a policy matter. The second question is more volatile and will lead to the most pushback. Should water intake from industry be limited only within the confines of the lake boundary, for instance, or do those limitations get pushed farther up the watershed to limit intakes upstream?

Applying the standard terms of western water law to this question, it would depend on when the water rights for the lake were deemed to be "perfected." Might they be deemed to have come into existence in the modern era—for instance, on the date that the conservation pool concept was enacted? In that case those water rights would take a back seat to any rights senior to them either within the lake boundary or upstream within the watershed. If those water rights were deemed to be in place at the time of statehood, however, when the lakebed was transferred to the state under the Equal Footing Doctrine, then all other water rights within the lake boundary or upstream within the watershed would be junior to those rights, so water withdrawals would have to be curtailed. I believe that the latter view is actually more defensible from a legal perspective, though it would be more controversial from the perspective of the water buffalos—a term used to describe powerful western water interests. But in the end the question will no doubt come down to whether we're willing to trade inefficient agricultural, industrial, and residential water practices for a healthy Great Salt Lake. That should prove to be an interesting discussion.

Notes

1. "Economic Significance of the Great Salt Lake to the State of Utah," Bioeconomics, Inc., January 26, 2012, 46, http://www.gslcouncil.utah.gov/docs/2012/Jan/GSL_FINAL_REPORT-1-26-12.PDF.
2. Robert W. Adler, "Toward Comprehensive Watershed-Based Restoration and Protection for Great Salt Lake."
3. For a discussion of the influences on inflow, see chapter 3 by Craig Denton and chapter 5 by Daniel Bedford herein.
4. "What about This Great Salt Lake?" Friends of Great Salt Lake, 1, http://www.fogsl.org/images/stories/2011/II-Phys.pdf.
5. For a further discussion of this issue, see chapter 3 by Craig Denton herein.
6. For a discussion of this issue, see *Utah v. United States*, 427 U.S. 461–62 (1976).
7. "Development of an Assessment Framework for Impounded Wetlands of Great Salt Lake," State of Utah, Department of Environmental Quality, Division of Water Quality, Salt Lake City, Utah, November 2009, 1–10.
8. See www.willardspur.utah.gov.
9. Wayne A. Wurtsbaugh, David L. Naftz, and Shane R. Bradt, "Eutrophication, Nutrient Fluxes and Connectivity between the Bays of Great Salt Lake, Utah," *Natural Resources and Environmental Issues* 15 (2009): 51.

10. For more on this, see State of Utah, Department of Natural Resources, Division of Wildlife Resources, http://wildlife.utah.gov/habitat/ans/phragmites.php.
11. "Economic Significance of the Great Salt Lake to the State of Utah," Great Salt Lake Advisory Council, January 26, 2012, 6.
12. "What about This Great Salt Lake?"
13. "U.S. Magnesium," http://www2.epa.gov/region8/us-magnesium.
14. See www.kennecott.com/library/media/NA--Free%20Visitors%20Center%20Days.PDF, 2.
15. "Public Notice," U.S. Army Corps of Engineers, http://www.spk.usace.army.mil/Portals/12/documents/regulatory/public_notices/200901213-NOI-PN.pdf.
16. "Authorization to Discharge under the Utah Pollutant Discharge Elimination System (UPDES)," Statement of Basis for KUC UPDES Permit UT0000051, State of Utah, Department of Environmental Quality, Division of Water Quality, Salt Lake City, Utah, 4, http://www.waterquality.utah.gov/UPDES/docs/2010/07Jul/KENNECOTT%20UTAH%20COPPER,%20LLC%20UT0000051.pdf; "Ground Water Quality Discharge Permit UGW350011 Statement of Basis," January 2011, 4, http://www.waterquality.utah.gov/GroundWater/gwpermits/kennecott-tailings/kennecott-tailingsSOB.pdf.
17. Utah Administrative Code Rule 317-2-14, table 2.14.2.
18. For more details, see "A Great Salt Lake Water Quality Strategy," State of Utah, Department of Environmental Quality, Division of Water Quality, Salt Lake City, Utah, 4, http://www.waterquality.utah.gov/greatsaltlake/index.htm.
19. Rule R-317-2, Standards of Quality for Waters of the State, State of Utah, Department of Administrative Services, Division of Administrative Rules, http://www.rules.utah.gov/publicat/code/r317/r317-002.htm.
20. "Golden Spike National Historic Site," United States Department of the Interior, National Park Service, http://www.cr.nps.gov/history/online_books/hh/40/hh40r.htm.
21. "Final Great Salt Lake Comprehensive Management Plan and Record of Decision," State of Utah, Department of Natural Resources, Division of Forestry, Fire, and State Lands, March 2013, http://www.ffsl.utah.gov/images/statelands/greatsaltlake/2010Plan/OnlineGSL-CMPandROD-March2013.pdf.

8

≈

The Colorado

Archetypal River

BROOKE WILLIAMS

IT BECAME CLEAR to me that cold day that I'd neglected most of what the Colorado River had to teach me. Sandy Ass Beach was beneath two feet of hard snow, flawless but for marks left by the sharp tips of dog claws and recent wind.

The winter of 2013 was mythically cold in the high desert, and weeks of sinking temperatures had frozen the Colorado River. Two or three times a week we drove the river's edge for fifteen miles between the turnoff to Castle Valley, where our home is, and Moab, where we buy books and food and gas. I use a large boulder near Big Bend to gauge the river's flow. Great Blue Herons fish at one of two particular turns in the river. Just out of town, Canada Geese are often seen wading in the shallows—along with mallards or mergansers, depending on the season. The color of the river changes daily based on the weather (from dark green to khaki to the color of persimmons moments after a flash flood in a side canyon). Winter hid all this. By mid-February I missed that river.

I worry about that river. The same climate changes that are causing monster storms and rising seas, famine, and fire are drying up the Colorado Plateau by starving the river that feeds it, and I needed to see it strong. I wanted it full.

Most of the year I take the dogs down to a small sandy cove that we call Sandy Ass Beach. Five years ago my friend Chris and I spotted the pure glowing sand of this Shangri-La from downstream and forced our way along the river's edge through dense willow and tamarisk. We knew we'd found a small paradise. Although occasionally we see signs that people boating the Daily (a popular stretch of river that commercial companies turn into a playground and locals use for regular pilgrimage) might take lunch breaks on our beach, we treat it as if it's ours. Somehow we own that beach. Some days the only tracks we find are those that we left the week before.

I'd never seen Sandy Ass Beach during such a hard winter. While the dogs tested the ice and checked their familiar willows and rocks, I squatted there chuckling about how in June I would dig a shallow hole and lie there naked. But the same cliff-shade that makes my beach bearable as the summer swelters and throbs had trapped and cooled the sharp, slow air, against which my down parka and wool hat were powerless that winter day.

Downstream there was no sign of the river, only a perfect surface like a white floor troweled smooth by a craftsman. Upstream the current compressed into a narrow channel had forced a black part of the river to the surface, where it curved between sensuous white snow banks as in a fairy tale. Directly in front of me lay total chaos. Foot-thick ice freed by the upstream current, meeting the immovable, had buckled and climbed and frozen into a disaster.

The Colorado River is the lifeblood of 30 million people. It waters 3 million acres of farmland and produces 1,800 megawatts of hydropower. It drains a quarter-million square miles in seven western states. For the Paiutes, Hopis, Utes, Navajos, Apaches, Ourays, Uintahs, Havasupais, and a dozen other tribes, the Colorado River is part of their historical legacy and continues to provide both spiritual and economic benefit. The river supports thousands of wild species, many endemic. Lifeblood, life force, this river is the archetype for this region, the Colorado Plateau, which for many is America's true heart.

Carl Jung, one of the great thinkers of all time, believed that archetypes are universal symbols shared by every human who ever lived, passed from generation to generation in a way similar to the way genes are inherited. The hero, the wise old man, the fool, the trickster, the Great Mother and sky father, the devil, the scapegoat, the warrior, the lover. All are archetypes present in the human psyche since we first appeared. Stones, caves, mountains, and rivers are also archetypes and are often transformational in nature. A cave appearing in a dream could mean death and rebirth, illumination; a stone could symbolize the simplest, deepest experience. Mountains are places of great insight. Rivers represent the passing of time or the boundary between one thing and another, between now and then. When archetypes appear in literature, they often elicit a deep emotional response.

Rivers are often characters in literature—such as the stages of life that the River Styx represents in the *Odyssey* or the freedom from captivity and child abuse and slavery that Huck Finn's mighty Mississippi represents. In *Siddhartha,* Hesse's river is a constant, foundational: "the river is everywhere at once, at the source and at the mouth, at the waterfall, at the ferry, at the rapids, in the sea, in the mountains, everywhere at once, and . . . there is only the present time for it, not the shadow of the past, not the shadow of the future."[1]

Jung suggested that archetypes are instinctual images, the symbols that represent the language of the collective unconscious, that "*layer where man is no longer a distinct individual, but where his mind widens out and merges into the mind of mankind . . . where we are all the same.*"[2] According to others, the collective unconscious contains the entire evolutionary history of our species, including all the tools that we've ever needed to save ourselves.

Jung again:

> Besides the intellect [we each have] symbols, which are older than the historical man, which . . . still make up the groundwork of the human psyche. It is only possible to live the fullest life when we are in harmony with these symbols. *Wisdom is a return to them.*[3]

As modern humans, we steer clear of this inner world. We've been seduced into thinking that all our technology and information and beliefs have made obsolete the vast amounts of evolutionary material existing beneath the conscious surface, beyond the ability of our science to explain and describe. We believe in a "God species,"[4] and that we are it. For most of our history this inner world was the source of our survival. And we've done very well. Could ignoring this "evolutionary material" be closing off most of the library in which our complete knowledge is contained? Due to the warming climate (the "super" storms, rising seas, and mass extinctions), the next decade may be the most important in the entire history of our species. The solutions that we have available to deal with our problems seem inadequate. Wouldn't this be a good time to use that old, deep part of the library (the places with dark wood, long tables with those iconic lamps, books without bar codes)? What do we have to lose? How do we enter that part of the library? Why is it closed? What are we afraid of?

Wisdom.

If wisdom is a return to our deep old symbols, we have forsaken it, selling it off to corporate interests who seem to have planned its obsolescence.

When we first moved to Castle Valley, I thought I had a deep relationship with the Colorado. I'd floated much of it on numerous river trips, both commercial and private. I was an above-average swamper—one of the people on most commercial trips who ties up the boat, coils ropes, deals with the groover (what river runners call the toilet) and serves as the sous-chef/dishwasher at meals. I'd hiked along its banks and explored many of its side canyons. Everything changed for me when I met the love of my life and abandoned my career path (I'd been hired to row the back-oar on a triple-rig in Cataract Canyon). Something shifted when this river became a regular part of our lives—daily we drive along it, swim in it, or stack its stones into art. Before, every trip to the Colorado required scheduling, planning—methodical thinking, conscious effort. Since moving close to the river, the planning and thinking and effort associated with it have all evaporated, along with any expectation. Now we just go. Like primitive hunters, we only know one thing: that we do not know what will happen next.

Something always happens.

Years ago Terry, Rio, and I were at the river late one hot summer day. We were floating in an eddy watching light drop and colors deepen, and just as the cliffs turned purple we saw the Hatch. A massive cloud of just-born mayflies suddenly engulfed us—mayflies in our eyes, hanging from our brows, mating in our hair. We breathed in mayflies and rinsed their nymph-skins off of our skin. Then it was over. I was of two minds. I wondered if it was just the situation or if they were a bit larger and lighter in color and perhaps of a different species than those with which I'd grown familiar. At the same time I thought about what this moment meant—how short life is, how fleeting, much of it taking place beneath all surfaces. Later my two minds joined on discovering the Latin name of the mayflies, *Ephemeroptera*, from the root "ephemera," meaning "all things that exist or are used or enjoyed for a short period of time."

Something always happens. Will I notice?

While I can count on one hand the number of mayflies I've seen since the millions I saw that summer evening, I've become obsessed with dragonflies and damselflies.[5] One bright spring day at Sandy Ass Beach I decided to learn some specifics about damselflies. I can pin it down to one individual—a damselfly with a red spot as if its metal wings were being welded to its body by a glowing torch. It was perched in a dead tamarisk. Years ago damselflies had become my totem animal when I was visited by one in a daydream (by this I mean the generosity with which they inhabit my life, the value and significance they bring to it). So powerful was this experience that it led me to consider the role that these creatures have played in the lives of people throughout history. In many cultures dragonflies and damselflies are the messengers between the inner and outer worlds and represent the souls of the dead. Since that moment the appearance of a dragonfly or damselfly has triggered a heightened awareness of what is going on around me. Either a dead relative is trying to communicate with me or I am receiving a message from the inner world. Or both.

The wild world has always fascinated me, and I consider myself a good amateur naturalist. I knew that there must be many dragonfly species, but the best I could do was put them into one of two categories. The larger dragonflies (usually metallic blue or green in color) hold their wings out perpendicular to their bodies when perched, while the smaller, blue damselflies lay their wings parallel along their back. That all changed the day the bright red damselfly, with its wings glowing red, perched in that tamarisk.

Back home that evening, I searched the Internet for clues about that damselfly, wings afire. Part of me hoped that I'd seen magic, the only being of its kind, a gift from the gods, a message. Part of me wanted to be exact, to learn its range, its habits, its long Latin name. No damselfly had ever mystified me as this one did.

American rubyspot (*Hetaerina americana*): even its photo was glorious.[6] I have eight books now—mainly field guides—on dragonflies and damselflies. I find that their metaphorical/archetypal significance deepens with my scientific knowledge, such as that gleaned from a recent article in the *New York Times*.[7]

I once spent fifteen minutes with a variegated meadowhawk dragonfly. I saw him from a distance and lay down in the water, inching closer and closer until I was within one foot of the tall blade of dried grass where he perched. I watched as he flew off eight times and returned seven times to chew and swallow the same species of small gnat.

Late that same summer I watched seventy-four pairs of powdered dancers (*Argia moesta*) on short logs floating in the shallow eddy. Males, attached to the back of the females' heads, hovered, watching, ready to pull their mates to safety should a predator approach while she released her eggs into the river. I sank into the mud while trying to photograph them and could barely pull myself out.

What happened to the mayflies, herons and geese, damsel eggs and nymphs in June 2011, when for a week the Colorado River flowed deeper and wider than I'd ever seen? Sandy Ass Beach was drowned.[8] Huge logs and live uprooted cottonwoods moved downstream like small islands, their limbs vertical with lime colored-spring leaves waving like flags. A massive late season snowstorm high in the headwaters was followed by weeks of sun and warmth that flooded the river to 90,000 cubic feet per second. Would the giant creature that this river had become, as it has done only rarely in the recent past, once again force its way across its desiccated delta to the sea?

While some river water may have trickled temporarily into the Gulf of California that summer, most of it was stolen by diversion canals or trapped by one of the many dams, though these lake levels have dropped dramatically in recent years. In 1922, when the Colorado River was divvied up, the region was experiencing an unusually wet period: the river's flow totaled 17.5 million acre-feet. Today the annual average of 14.7 million acre-feet must be allocated among seven states and Mexico, all of which are experiencing exploding urban populations and climate change. Recent studies suggest that a warming climate will affect the Colorado Plateau more than the rest of the country. Hotter, drier weather could reduce river flows by 10 to 30 percent. This host of challenges led to the American Rivers listing of the Colorado as America's Most Endangered River in 2013.[9]

The bigger question for me was what will happen to the mayflies and the dragonflies and damselflies when drought diminishes the river? What will happen to the willows and the willow flycatchers and the fish and the ospreys that hunt them? What will happen to the evening light with less to reflect it?

The next year, on one of the last autumn days, the river seemed low enough to walk across. I thought of a recent photograph of the riverless Colorado River delta, littered with old tires and plastic bottles and oozing with brown foaming muck. For tens of thousands of years those were healthy wetlands, loud with the songs of birds and the chatter of insects, a place soothed by the soft sounds of reeds and marsh grasses rubbing against one another. Before dams and droughts, dolphins could be seen catching the fish that they'd chased sixty miles up the river. They say that the incoming tide racing up the river made "a noise like a train."[10]

The newly exposed slime at the base of Sandy Ass Beach gave off fetid, organic smells. Imagining the fumes rising from what the Colorado River Delta has become frightened me. The river was so low that a new island appeared upstream from Sandy Ass beach. The air temperature was so perfect that it disappeared. The dogs lay in the sand next to me, watching the river flow past. I too became a resting dog—just watching. An autumn leaf fell from an oak on the bank above us. They watched it. I watched it. A small fish rose in the eddy. We all watched it. Then above that new grassy island something began to happen. Hundreds of damselflies—I could only see them when low light from the setting sun came in at the perfect angle—were feeding on some invisible late season hatch.

What happens to us when our archetypes lose their source? What happens when we dream of an animal that has gone extinct or the passing time symbolized by a river that now quits before it ends? Will the changing climate and the drought associated with it force us to create new archetypes? (I would have missed those feeding damselflies had the river not dropped to that depth.)

Then came this cold winter and the only sunny days in the past week. As we always do, we pull off the road. The dogs are crazy to get out. The moment the door opens, they're down the trail, now buried beneath snow. As I grab my pack and close the door, they have already disappeared. Theirs are the only tracks in the snow, barely softened by the afternoon sun. I follow them, turning east and passing by the large, turtle-shaped boulder (long ago we pioneered a new way to Sandy Ass Beach, completely avoiding the river's complex edge) then into and out of the wash and up the hill where they've just passed the three giant cottonwoods standing like statesmen below. Although the trail is invisible, the dog track follows it perfectly between the willows and the squawbush. I've forgotten which of the three tunnels through the tamarisk leads to Sandy Ass Beach, and that old pull-top beer can that usually guides me has disappeared in the snow. No matter, the dogs know the way.

Half an hour later, sitting hunched in the snow on that cold day, stiff but not quite paralyzed, I watch the river blossom into its archetype.

Downstream the frozen white floor of river ice is covered by soft snow—I was there once, oblivious to the life force flowing beneath me. Everything I thought I needed was on that surface. Anything beneath it seemed old, antiquated, at best inconsequential and at worst distracting. That was the past. Now, in the present, with the ice directly in front of me, I know that the river is down below, inside, flowing strong and purposeful, deep and green and muscular and filled with stories and clues. On top the ice chunks buckling and bending and twisting before breaking loose become my own chaotic transformation as the deep river forces its way through the surface from below. Upstream, tomorrow or next month or next year, the river's strongest current will move quietly, part of the surface. The inner and outer rivers (my inner and outer worlds) will soon join into one.

Notes

1. Hermann Hesse, *Siddhartha*, 1172–74.
2. Quoted in Edward Hoffman, *The Wisdom of Carl Jung*, 60 (emphasis added).
3. Ibid., 22 (emphasis added).
4. Mark Lynas, *The God Species: How the Planet Can Survive the Age of Humans.*
5. Dragonflies have front wings that are broader than their hind wings and hold their wings out at rest; damselflies have front and back wings that are the same size and are held over their body (parallel to the body axis) at rest.
6. See http://www.cirrusimage.com/damselfly.htm for an image.
7. Natalie Angier, "Nature's Drone, Pretty and Deadly," *New York Times*, April 1, 2013, http://www.nytimes.com/2013/04/02/science/dragonflies-natures-deadly-drone-but-prettier.html?_r=0.

8. See http://www.youtube.com/watch?v=Z2At_vLhq5s to view a video that I made of that day.
9. American Rivers, http://www.americanrivers.org/; see chapter 10 by Annette McGivney herein for more discussion of issues facing the Colorado River.
10. Henry Fountain, "Relief for a Parched Delta," *New York Times*, April 16, 2013, http://www.nytimes.com/2013/04/16/science/earth/optimism-builds-for-effort-to-relieve-a-parched-delta-in-mexico.html?pagewanted=all&_r=0.

9

≈

Going with the Flow

Navigating to Stream Access Consensus

SARA DANT

The Sunday Drive: an idyllic American pastime from a bygone era. On a fine May morning in 2012, my daughter and I set out to revive this ritual and continue a family tradition begun by my grandparents, whose love of cars flows deep in my veins. For them, the Sunday Drive was a conspicuous parade that demonstrated on a very basic level that they were thriving... managing to stay above the rising tide of the Great Depression and keeping up with the Joneses very nicely, thank you. Auto mania must skip a generation, however, because their eldest son, my father, believes that cars are primarily a practical means of conveyance. He had a wife and young daughter, so he bought a Buick—big, safe, green. It floated along the desert boulevards of my Phoenix youth with a heaviness that projected stability (despite none of us wearing seatbelts). Safety and reliability are my priority with my own eight-year-old, so on this beautiful Wasatch spring morning we are humming along in a six-speed manual Volkswagen clean diesel, enjoying forty-five miles per gallon. At least it's fire-engine red.

After a dark, cold winter cooped up indoors, a celebration of longer, warmer days and the biological riot of erupting greenery seemed in order. I was researching the history of the Weber River at the time, so my daughter and I decided that a field trip along this meandering waterway was just the ticket. We began in Ogden, just upriver from where the Weber joins the Ogden River before their co-mingled waters finally delta into the Great Salt Lake a short distance later. From here we headed east through the mighty crack in the Wasatch carved by the river, on I-84, past the lush,

Weber-irrigated fields of aptly named Mountain Green, and on up through the pastoral heart of Summit County, with the winding water as our constant companion.

For the entire length of our journey, my daughter was completely enthralled—no electronic gadgets distracted her. She, like me, was on a quest to know this river in all of its glory. As a good eight-year-old she fully expected that our adventure would be both scenic and tactile. People don't drive nearly 200 miles along a rushing, bouncing freshet only to admire it from the other side of a car window. No, a river is to be experienced, preferably with toes. We were both in for a rude awakening. "No Trespassing!" "Keep Out!" "Violators Will Be Prosecuted!" and barbed wire greeted us at every possible access point along the upper stretches of the river.

"No fair, mom!" was her response. The girl had a point. Our sole objective that day was to splash our feet in and skip a few rocks across this artery that had once conveyed ties and timbers for the mighty transcontinental railroad and the booming mines of Park City. To be sure, the lands on either side of the river were privately owned, obviously by wealthy landholders and resorts, but we weren't interested in their land. We weren't planning to camp or build a fire or stake a claim. We were simply trying to enjoy the river that had been so essential and vital to the early pioneers of the later Beehive State. That, evidently, was too much to ask. Although I managed to salvage the Sunday Drive adventure with a mint-chocolate-chip milkshake in Kamas, the question of stream access—why we *couldn't* access the stream—haunted me, as did my daughter's lament.

As it turns out, the question of fairness lies at the heart of Utah's stream access debate. Residents of this arid region have never shied away from water wars. In addition to issues involving irrigation and drinking water, Utahns have recently grappled with questions regarding recreation access to the state's rivers and streams, like the Weber, that cross private property boundaries. It is serious business. Sport-fishing alone pumps about $259 million into Utah's outdoor recreation economy each year.[1] Unfortunately, as is so often the case with conflicts involving natural resources, meaningful dialogue has deteriorated, legal challenges are flying, and rival factions—private property owners and recreation advocates—have each come to see the other as the "enemy." Yet the matter of stream access, like many other recreation conflicts, actually has a reasonable center. If the debate's participants are willing to cooperate, a consensus solution could not only save countless hours of courtroom battling and hundreds of thousands of litigation dollars better spent on stream restoration and maintenance but also provide a sustainable long-term solution to managing the state's water and riparian resources for the good of the people, all the people.

While the topic of stream access has been in the news and the law for many years, the issue blew up again in March 2010, when the Utah State Legislature passed and Governor Gary Herbert signed into law a new stream access bill known as HB 141. Also called the "Public Waters Access Act," HB 141 had ignited controversy from the moment Representative Kay McIff (R) introduced it in the legislative session, because it constricted the broader recreation rights conveyed in earlier court decisions by recognizing only a "limited recreational floating right" through private property. Essentially the law prohibits and criminalizes all public recreational use of rivers and streams that cross private property except floating and the "incidental touching" of stream beds and banks necessary to continue floating. McIff's concern was that the court's guarantees of public access to waterways had trumped private property rights. HB 141 inverted that paradigm by making private property rights primary. As the law specifically states, "the Legislature shall govern the use of public water for beneficial purposes, as limited by constitutional protections for private property."[2]

Public use advocates immediately cried foul, claiming that the new law was "a motivation for people not to fish in Utah."[3] And they wasted little time mounting their opposition. By November 2010 the pro-access Utah Stream Access Coalition (USAC) had filed suit challenging the new law on the grounds that it violated the public's constitutional ownership and use of these waterways and the state's obligation to preserve and protect that ownership and right of access.[4] While the specific target of the USAC's ongoing (as of 2014) lawsuit is the Victory Ranch, a bucolic, sprawling development bisected by the blue-ribbon Provo River above the Jordanelle Reservoir, the USAC's larger goal is to overturn HB 141.

In 2011 the USAC filed a second stream-access lawsuit for the Weber River. In this case the USAC argues that in the late 1800s Utahns used the upper Weber River, like many other northern Utah waterways, to transport railroad ties, logs, and other wood products to market, making the river "navigable" under federal law at the time of statehood. If this suit is successful, HB 141—a state law—would not apply and the public would own both the bed and the waters of the Weber River reach at issue.[5] Moreover, it would establish a legal precedent for determining the navigability of other Utah rivers and streams.

As these lawsuits wend their way through the courts, they necessarily lead to broader questions about stream access in the state. The questions of who gets to use the state's waters and which waters are navigable demand close historical examination, in terms of both the actual uses of the state's streams and rivers over time and the laws that have attempted to define public and private rights. Ideally, a more thorough understanding of the past will lead to the development of thoughtful and fair solutions in the present.

When the Saints first arrived in Deseret in 1847, Mormon leader Brigham Young, and the nascent LDS Church, promoted communalism and governing policies that assured public access to and use of the region's streams, rivers, wood, and timber as part of his "land law."[6] In 1852 this ideal was codified in territorial law, which proclaimed that the county courts had "control of all timber, water privileges, or any water course or creek, to grant mill sites and exercise such powers as in their judgment shall best preserve the timber, and subserve the interest of the settlements, in the distribution of water for irrigation, or for other purposes."[7] Indeed, as a later irrigation economist argued, "that no monopoly in either land or water developed in the early days was due to the fact that the church leaders were constantly on guard against it."[8]

In September 1852, armed with these ecclesiastical and legal blessings, Robert Gardner and his brother made the first recorded commercial survey and assessment of the "weaber [*sic*] for timber and floating purposes." He "found the River good for floating" and also noted "some beautiful land and extensive range for stock" in the canyon. The men found similar favorable conditions along the Provo River, which Robert Gardner characterized as "nearly as large as the Weber" and "as handsome a stream for floating purpose as could be desired." Gardner's diary stated that both rivers had sufficient flows to sustain a saw mill, "plentiful" fir trees, and flows "large enough in times of high water to float timber from points many miles back in the mountains."[9] Archaeologist James Ayers adds that in the Uintas (for example, Bear and Black's Fork Rivers), "1850s and early 1860s logging activity was sporadic and of a relatively minor nature."[10]

Nevertheless, Gardner's reconnaissance presciently foretold the future uses of the Weber and other rivers of northern Utah as highways of commerce and trade.

Public use of and access to the region's rivers continued through statehood. By the 1850s, according to a United States Department of Agriculture report and various secondary sources, settlers along the lower course of the Weber in places such as Uinta (aka Easton and East Weber) and along the Provo River had established more than 100 sawmills to process logs cut up in the canyons and floated down the rivers.[11] Similar activities occurred on the Ogden River to the north.[12] One biography of Ogden pioneer Lorin Farr, the city's first mayor, details how groups of men would ascend Ogden Canyon in search of suitable timber. They would fell the trees, mark and cut them into logs, and then "float them down the Ogden River for retrieval"—a process "particularly effective during spring floods of the river."[13] This use of the river was economical for a number of reasons, but it also allowed sawyers to avoid the rather steep tolls assessed on roads such as that through Ogden Canyon, where travelers during the 1860s paid a dollar for a loaded wagon, fifty cents for an unloaded wagon, and a quarter for mounted horseback passage.[14] In nearby Morgan County sawyers felled logs and hauled them by oxen or floated them down the river to the sawmills.[15]

Ayers also documents similar timbering/floating activities on the Bear River: as a result of the transcontinental railroad, "beginning in 1867 and continuing through 1869, the most intensive exploitation of the area took place."[16] Even though some men lost their lives in this dangerous occupation, the timber industry provided many jobs and was central to Utah's economy during this period.[17] In fact, the tie-driving phenomenon was common throughout the Rocky Mountains. Typically 30–100 men worked the larger drives, which commenced when the rivers began to rise in late spring.[18] In 1872 the Colorado Territorial Legislature even legally protected the public access and use rights of log drivers to use the rivers, stating that it was "lawful for any person or persons to float any and all kinds of timber, such as saw logs, ties, fencing poles or posts, and firewood, down any of the streams of this Territory" so long as they posted a bond sufficient to cover any damages.[19] Downstream booms, often at the conjunction of the river and the railroad, captured the drive, which was then loaded onto wagons or railroad cars.

Although the Union Pacific and Central Pacific finally joined on May 10, 1869, in Utah, the need for public stream access remained.[20] In northern Utah, the setting for our Sunday Drive, the 1870s saw the beginning of the Park City mining boom, the construction of the Summit County narrow-gauge railroad between Echo and Coalville (purchased from the LDS Church by the Union Pacific in 1877), the expansion of the Utah Southern Railroad, and a proposed Utah Eastern Railroad from Coalville to Salt Lake (to provide competition to the Union Pacific in the so-called coal wars).[21] The Union Pacific spur and a narrow-gauge track built by the Utah Central both passed through Wanship on their way to Park City, acting as a conduit of coal, timber, and silver. Park City fortuitously boomed as a mining town at precisely the same time that the transcontinental railroad connected. All of these endeavors required large infusions of timber, which the Weber Canyon and its river readily provided.[22] In her autobiography, Olive Emily Somsen Sharp describes the work of her father (Henry J. Somsen) in the 1870s "getting timber from the high mountains east of Coalville and Kamas, Utah": "the men worked from the Provo River, north to the Weber River, where they floated the ties down the Weber to Echo."[23] This account is corroborated by the July 16, 1877, edition of the *Deseret Evening News*, which reported that "railroad ties that have been cut in East Weber Mountains are being floated down the Weber River in large numbers to Echo."[24]

The Weber and other rivers in northern Utah were known not only for their economic utility but also for their prime angling. Angler angst had in fact been the genesis of the USAC's lawsuits. Beyond the working nature of the Weber and other rivers was their historic value to local sport fishers, who regularly waxed eloquent in praise of the "piscatorial pleasures" and the fine "finny favorites" along the Weber River that would "answer magnanimously to the angler's wiles."[25] A colorful travel narrative of his 1877 passage along the Union Pacific penned by Robert E. Strahorn, for example, describes the "unexcelled trout-fishing in Bear river," "the gamiest of trout" in the Ogden River, and "first class" trout fishing on the Logan River, in addition to detailing that "over 1,000,000 feet of timber, and some 200,000 railroad ties were cut from the neighboring mountains in 1878. Ties are floated down the Bear river, thence down to the Pacific Railway."[26] Strahorn's narrative also describes tie drives throughout the region, indicating the common nature of the practice. Furthermore in 1879, in addition to the Weber and other rivers' tie drives, the Provo City

Semi-Weekly Enquirer reported that "the railroad tie business is assuming massive proportions here. Thousands are being floated down the Provo River." The *Logan Leader* estimated that "between 100,000 and 200,000 ties have been floated down Logan river this season."[27] The *Logan Leader* also recorded that tie contractors "Coe and Carter have spent about $60,000 here this season."[28]

An 1888 "enormous" and "immense" tie drive on the Provo River was certainly lucrative—more than one newspaper article indicated that it covered eighteen miles of river, contained at least 100,000 ties, and would "put about $50,000 in circulation in Provo"—but it was also problematic: "there is some trouble anticipated in regard to interference with irrigation."[29] The tie and log drives down the Weber and other rivers may initially have been economical and exciting, but by the end of the century they were also controversial. Farmers increasingly complained that the rafts of logs careening down the rivers destroyed delicate irrigation systems and caused excessive damage.[30] As early as 1882 the Provo City Council received complaints that "the city damns [*sic*] had been damaged by parties floating ties down Provo river." In June 1888 none other than future Utah senator (and soon-to-be member of the LDS Church Quorum of the Twelve Apostles) Reed Smoot appealed to the Provo City Council to stop the "heavy drive of ties down the Provo River this month," out of concern that it would "cause great damage to irrigation and machinery interests by disturbing the bed of the river, destroying dams, etc."[31] In other words, what had once been a cost-effective and efficient use of the rivers had sometimes become destructive and expensive.

Smoot was a powerful rival for the tie drivers, and by 1890 his concerns had become territorial law: anyone "who shall raft or float timber or wood down any river or stream of this state and shall allow such timber or wood to accumulate at or obstruct the water gates owned by any person or irrigation company...is guilty of a misdemeanor."[32] It is important to note here, however, that the dispute was not an attempt to privatize the rivers but rather to preserve their "greatest good for the greatest number" function. So long as tie drivers were conscientious and careful and did not destroy others' property, their rights to use the river for commerce were preserved.

By the turn of the twentieth century much of the work being done by the Weber and other rivers along the Wasatch and Uinta ranges had shifted from primary use as a conduit of timber to providing irrigation and municipal water supplies.[33] By 1903, for example, the number of canals and ditches drawing water from the Weber River had ballooned from about a hundred at the time of statehood to more than one hundred and fifty.[34] The railroad's ability to haul in cheap timber from Oregon also cut into the tie business in the region.[35] Recreation in the form of camping and fishing expeditions continued to provide entertainment, respite, and an economic boost to local economies, however, and the state fish and game department regularly stocked the Weber River and its tributaries to encourage this public use.[36]

Recreational interests, the historical predecessors of the USAC, had their own concerns about the Weber River. In 1910, for example, one report from the July 9 edition of the *Deseret News* quoted a thwarted fisher who argued that "the fishing is poor and I ascribe this mostly to the dynamiters. Between Rockport and White's I heard seven shots in two days." The irate fisher went on to complain that he had paid $1.25 for his fishing license: "the non-enforcement of state laws" constituted a "disgrace."[37] Faced with this kind of biting criticism, the state's fish and game commissioner stepped up efforts to regulate access to the state's waters. In June 1920, R. H. Siddoway reminded anglers "to remember that property owners, through whose lands the fishing streams run, have rights which should be recognized… [and] to keep as closely as possible to the channels of the streams." Fishers, he continued, "have rights also." Echoing the early sentiments of Brigham Young, Siddoway stated that "the waters of the state belong to the state. The fish therein are also the property of the state. Fishermen may wade any of the streams of the state. If ordered off the property of any owner thereof, they cannot be ordered out of the streams."[38]

Legal discussion of stream access began even before Utah became a state and co-evolved with the historic uses discussed above. In 1787 the Confederation Congress (under the national Articles of Confederation government) passed the Northwest Ordinance, which set important precedent for the future state of Utah by stating that "the navigable waters leading into the Mississippi and the Saint Lawrence, and the carrying places between, shall be common highways, and forever free,… without any tax, impost, or duty therefor." The attempt to balance the right of public access and private property rights appeared in various early Utah territorial laws: in 1852 and 1890 (as discussed earlier in this chapter) and also in 1888, when legislature decreed that "all navigable rivers, within the Territory occupied by the public lands, shall remain and be deemed be public highways."[39] In 1896, when Utah achieved statehood, the constitution affirmed that "all existing rights to the use of any of the waters in this State for any useful or beneficial purpose, are hereby recognized and confirmed."[40]

Subsequent court cases sought to clarify public and private rights and protections. In 1937 the state's high court waxed eloquent when it declared that Utah's "waters are the gift of Providence: they belong to all as nature placed them or made them available...While it is flowing naturally in the channel of the stream or other source of supply, [water] must of necessity continue common by the law of nature...or property common to everybody. And while so flowing, being common property, everyone has equal rights therein or thereto, and may alike exercise the same privileges and prerogatives in respect thereto."[41]

In 1982 the state Supreme Court more directly addressed the public access question when it ruled in *J.J.N.P. Co. v. Utah* that "the State regulates the use of the water, in effect, as trustee for the benefit of the people. Public ownership is founded on the principle that water, a scarce and essential resource in this area of the country, is indispensable to the welfare of all the people; and the State must therefore assume the responsibility of allocating the use of water for the benefit and welfare of the people of the State as a whole." The ruling further argued that "a corollary of the proposition that the public owns the water is the rule that there is a public easement over the water regardless of who owns the water beds beneath the water. Therefore, public waters do not trespass in areas where they naturally appear, and the public does not trespass when upon such waters." Thus, the court concluded, "private ownership of the land underlying natural lakes and streams does not defeat the State's power to regulate the use of the water or defeat whatever right the public has to be on the water. Irrespective of the ownership of the bed and navigability of the water, the public, if it can obtain lawful access to a body of water, has the right to float leisure craft, hunt, fish, and participate in any lawful activity when utilizing that water."[42]

Finally, in 2008 the Utah Supreme Court issued its opinion in *Conatser v. Johnson*. The decision relied heavily on the precedents set by *J.J.N.P. Co. v. Utah* and established that "the scope of the public's easement in state waters provides the public the right to engage in all recreational activities that *utilize* the water and does not limit the public to activities that can be performed *upon* the water" (emphasis in the original). The court then affirmed public easement rights to touch even privately owned river beds: "The practical reality is that the public cannot effectively enjoy its right to 'utilize' the water to engage in recreational activities without touching the water's bed." But the court was also mindful of private property interests, writing that "[t]he right of the easement owner and the right of the landowner are not absolute, irrelative, and uncontrolled, but are so limited, each by the other, that there may be a due and reasonable enjoyment of both…the public may not cause unnecessary injury to the landowner. If the public acts beyond these strictures, it has exceeded the scope of the easement."[43] The court argued for fairness and balance, in other words.

For private property owners, however, *J.J.N.P. Co.* and *Conatser* left several critical questions unanswered: owner liability, fencing issues, and the extent of the "bed" (highest high-water mark or "normal" high-water mark?) among others. Private property advocates also argued that the courts' rulings provided a significant *disincentive* for riparian rehabilitation and maintenance: "public access to environmental resources promotes overuse, which reduces environmental quality. Private ownership, on the other hand, promotes good resource stewardship because owners capture the benefits of their investments."[44] This brings us full-circle back to HB 141.

In early 2013 hopeful signs of cooperation peeked through the frosty contention like the first bulbs of spring along the Wasatch. In February, sponsors of rival state house bills regarding public stream access agreed to withdraw their proposals and allow the judicial process to run its course. As state representative Dixon Pitcher (R) acknowledged, stream access proponents "are only asking for a compromise. They understand the importance of property rights but at the same time that the water belongs to the public."[45] In March access advocates got a major boost from the state's Fourth Judicial District Court when its summary judgment found in favor of the USAC on several counts. The court's opinion argued that the state had an obligation "to protect not only the 'the traditional triad' of public trust rights—navigation, commerce, and fishing—but also the ecological integrity of public lands and public recreational uses." And while "the State Legislature has authority to assess the relative importance of competing water uses…the Legislature cannot have undivided loyalty to one stakeholder in public lands. Rather, it must 'consider the health, safety, and welfare of all [Utah citizens]' and has a duty to 'manage and preserve public lands for present and future generations.' There is a presumption that in balancing these often competing interests, the Legislature acts for the benefit of the people, and not to serve private interests."[46]

Despite this pro-access rhetoric, the district court stopped short of overturning HB 141. Although the summary judgment stated that "the Act [HB 141] so narrowly defines lawful access to public water as to be tantamount to disposition of the public's interest," it also affirmed the right of the legislature to regulate both the use of and recreational easement on public waterways. The judge concluded his ruling by asking both sides to submit additional briefings regarding the state's public trust doctrine before he issued his final judgment. In the spring of 2014, a legislative "Compromise Bill" (HB 37) that would have led to the dismissal of both court cases failed to escape the House Rules Committee and batted the issue back to the courts.

As both sides double-down in their efforts to find a just, fair, and sustainable solution to the question of stream access in Utah that protects both public access and private property, they would be wise to study their water history and revisit the thoughtful counsel of R. H. Siddoway, the state's fish and game commissioner from 1917 to 1921. He reminded anglers and property owners, in the language of the day, that the best way to ensure the rights of all was to "be a gentleman."[47]

In the end private property owners have the right to expect and protect the sanctity of their holdings, to be sure, but the people also have the right to access and enjoy their public waters, as they have throughout history. In arid states like Utah water acts like a magnet, and its relative scarcity only enhances its perceived and monetary worth. But this precious commodity ought to be available to all, not just to the highest bidder. Surely there is a middle ground that allows property owners to maintain the integrity of their lands while still providing right-of-way access to wading anglers and splashing eight-year-olds in the rivers that run through them. It's only fair.

Notes

1. Man-Keun Kim and Paul M. Jakus, "The Economic Contribution and Benefits of Utah's Blue Ribbon Fisheries," November 5, 2012, Department of Economics, Utah State University, copy in author's possession. See also Brett Prettyman, "USU Study Says Fishing Nets $259 Million for Utah Economy,"

Salt Lake Tribune, April 19, 2013, http://www.sltrib.com/sltrib/money/56138236-79/blue-ribbon-anglers-utah.html.csp.

2. Kay L. McIff, "Recreational Use of Public Water on Private Property," Enrolled Copy HB 141, http://le.utah.gov/~2010/bills/hbillenr/hb0141.pdf.

   In Utah the question of stream access is largely a legal access issue. In other western states, however, private landowners often band together to make the public's actual physical access to legally navigable streams difficult, which in turn sets off a whole other set of legal battles.
3. Matthew Frank, "Montana's Stream Access Law Stays Strong."
4. Christopher Smart, "Utah Group Asks Court to Rule on Stream Access Law." For a detailed discussion of USAC's position, see Craig C. Coburn and Kallie A. Smith, "Memorandum in Support of Utah Stream Access Coalition's Motion for Summary Judgment," September 2, 2011, http://www.i9studios.com/USAC/LegalDocs/USAC_v_ATCRealtySixteen_MemorandumInSupportOfMotionForSummeryJudgment_TS8631.pdf.
5. HB 141 *does* state that "[t]he public may use a public water for recreational activity if the public water is a navigable water." See McIff, "Recreational Use of Public Water."
6. B. H. Roberts, *A Comprehensive History of the Church of Jesus Christ of Latter-day Saints*, 269. See also Leonard J. Arrington, Feramorz Y. Fox, and Dean L. May, *Building the City of God: Community and Cooperation among the Mormons*, 57.
7. Section 39, Territory of Utah Legislative Assembly, *Resolutions and Memorials Passed by the First Annual, and Special Sessions, of the Legislative Assembly of the Territory of Utah, Begun and Held at Great Salt Lake City, on the 22nd Day of September, A.D. 1851 also the Constitution of the United States, and the Act Organizing the Territory of Utah* (Salt Lake City: Brigham Young, 1852), 46. See also John Swenson Harvey, "A Historical Overview of the Evolutions of Institutions Dealing with Water Resource Use and Water Resource Development in Utah, 1847 through 1947" (MS thesis, Utah State University, 1989).
8. Wells A. Hutchings, *Mutual Irrigation Companies in Utah* (Logan: Utah Agricultural Experiment Station, 1927), 15.
9. Robert Gardner, "Robert Gardner Journal, 1852 September," MS 6063, Church History Library, Church of Jesus Christ of Latter-day Saints, Salt Lake City (hereafter Church History Library).
10. James E. Aycrs, "Historic Logging Camps in the Uinta Mountains, Utah," 251; James E. Ayers, "Standard Timber Company Logging Camps on the Mill Creek Drainage, Uinta Mountains, Utah"; see also "Early Days in Ogden," *Deseret Weekly*, February 23, 1895.
11. David J. Farr, "Biography of Lorin Farr—Part 16," http://winslowfarr.org

/biographies/LorinFarr/lf16.htm; Mark E. Stuart, "Uintah," in *Utah History Encyclopedia*, http://www.uen.org/utah_history_encyclopedia/u/UINTAH.html. Forest Service Intermountain Region, U.S. Department of Agriculture, *Forest and Range Resources of Utah: Their Protection and Use* (Washington, DC: U.S. Department of Agriculture, 1930); Milton R. Hunter, ed., *Beneath Ben Lomond's Peak: A History of Weber County, 1824–1900* (Salt Lake City: Publishers Press, 1966), 169; Mark E. Stuart, "A Brief History of the City," http://www.uintahcity.com/history.htm; see also Alfred Lambourne, "Alfred Lambourne Writings, circa 1912," MS 4110, Church History Library; Hunter, *Beneath Ben Lomond's Peak*, chapter 12; Charles S. Peterson and Linda E. Speth, *A History of the Wasatch-Cache National Forest*, September 25, 1980, http://www.fs.usda.gov/Internet/FSE_DOCUMENTS/stelprdb5053310.pdf, 111–21; see Douglas M. Bird, "A History of Timber Resource Use in the Development of Cache Valley, Utah" (MS thesis, Utah State University, 1964), 58–64, for a list of sawmills in Cache Valley.

12. Farr, "Biography," 12.
13. Ibid.; see also F. Ross Peterson and Robert E. Parson, *Ogden City: Its Governmental Legacy—A Sesquicentennial History*, 31; Hunter, *Beneath Ben Lomond's Peak*, chapter 13; Richard C. Roberts and Richard W. Sadler, *A History of Weber County*, 125; Ralph B. Roberts, "Sawmills," 1944, Special Collections and Archives, Utah State University, Logan, Utah; Peterson and Speth, *A History of the Wasatch-Cache National Forest*, 121.
14. Roberts and Sadler, *A History of Weber County*, 125.
15. See, for example, Linda H. Smith, *A History of Morgan County*, 188; Elnora Arave Cox and Frederick James Wadsworth, *Biography of Abiah Wadsworth*, December 21, 1979, http://www.leavesonatree.org/histories/Biography%20of%20Abiah%20Wadsworth.pdf; Mrs. William Chadwick Stoddard, "History of Morgan County," *Morgan County News*, May 2, 1947.
16. Ayers, "Historic Logging Camps," 251; Ayers, "Standard Timber Company"; see also L. J. Colton, "Early Day Timber Cutting along the Upper Bear River"; and William Harvey Wroten Jr., "The Railroad Tie Industry in the Central Rocky Mountain Region: 1867–1900" (PhD diss., University of Colorado, 1956), 14.
17. Colton, "Early Day Timber Cutting," 203–4; see also Wroten, "The Railroad Tie Industry," 11–15, 75, 218; Olive Emily Somsen Sharp, "Autobiography," MS A 2038, Utah State Historical Society, Salt Lake City, 4; Michael K. Young, David Haire, and Michael A. Bozek, "The Effect and Extent of Railroad Tie Drives in Streams of Southeastern Wyoming," 126; Peterson and Speth, *A History of the Wasatch-Cache National Forest*, 125–26; Colton, "Early Day Timber Cutting," 204 (photo); and Robert E. Gresswell, Bruce A. Barton,

and Jeffrey L. Kershner, eds., *Practical Approaches to Riparian Resource Management: An Educational Workshop* (Billings, MT: U.S. Bureau of Land Management, 1989), 189; for a colorful, fictionalized version of tie hacking in the High Uintas, see Roy Lambert, *High Uintas Hi!*

18. Wroten, "The Railroad Tie Industry," 270.
19. Ibid., 280.
20. Ibid., 45, 49, 50n137; "History of the Cache National Forest" (unpublished paper, 1940), Special Collections and Archives, Utah State University, 12; see Brad Hansen, "Tie Drives in the Bear River Range" (unpublished paper, April 25, 2012), copy in author's possession, for details on the Bear River; Thomas X. Smith, "Account Book, 1879–1881," MS 11241, Church History Library, for details on the Logan River.
21. Don Strack, "Utahrails.net," utahrails.net; Edward L. Sloan, ed., *Gazeteer* [*sic*] *of Utah and Salt Lake City*; Brigham Young, "Office Files, 1832–1878," CR 1234-1, Box 104, folder 22, Church History Library; Robert E. Strahorn, *To the Rockies and Beyond* (Omaha: New West Publishing Company, 1879), 93–94; "Park City Items," *Salt Lake Weekly Tribune*, June 5, 1880; H. L. A. Culmer, ed., *Utah Directory and Gazetteer for 1879–1880* (Salt Lake City: H. L. A. Culmer and Co., n.d.); George E. Pitchard, *A Utah Railroad Scrapbook* (Salt Lake City: George E. Pitchard, 1987).
22. See, for example, Charles S. Peterson, "'Book A—Levi Mathers Savage': The Look of Utah in 1873," *Utah Historical Quarterly* 41, no. 1 (Winter 1973): 9.
23. Somsen Sharp, "Autobiography," 5.
24. *Deseret Evening News*, July 16, 1877; see also Peterson, "'Book A,'" for discussion of sawmill life along the Weber.
25. "Fishing," *Ogden Herald*, June 15, 1887; see also A. S. Condon, "A Big Wheel," *Ogden Herald*, May 16, 1887.
26. Strahorn, *To the Rockies and Beyond*, 88, 93, 122.
27. *[Provo City] Semi-Weekly Enquirer*, May 24, 1879; "Local Lines," *Logan Leader*, October 30, 1879; see also *[Provo City] Semi-Weekly Enquirer*, August 6, 1879; and Smith, "Account Book," for similar information on the Logan River; and Linda Carter, "Tie Drives Down the Provo River: Consolidated Sources List," n.d., copy in author's possession, for Sevier River.
28. Bird, "A History of Timber Resource Use," 33.
29. "Provo Points," *Salt Lake Herald*, June 17, 1888; "Random References," *[Ogden, Utah] Standard*, June 16, 1888; "Tie Drives in Provo River," *[Ogden, Utah] Standard*, May 20, 1888; "The Tie Drive," *Park [City, Utah] Record*, June 23, 1888; "Provo Points," *Salt Lake Herald*, July 22, 1888; "Floating Fragments," *[Utah] Daily Enquirer*, June 26, 1888.
30. See, for example, Wroten, "The Railroad Tie Industry," 280.

31. Carter, "Tie Drives," 7, 8; "City Council," *Utah Enquirer*, June 1, 1888 (quotations); *Deseret News*, June 13, 1888.
32. "An Act to Protect Irrigation Companies," March 11, 1890, in *Laws of the State of Utah, 1890–94* (Salt Lake City: Star Print Co., n.d), http://babel.hathitrust.org/cgi/pt?id=uc1.b3830829;view=1up;seq=2; see also "That Tie Drive," *Utah Enquirer*, June 8, 1888.
33. See R. L. Polk, *Utah State Gazetteer and Business Directory*, 1st ed. (1900), 979.205 U89G, Church History Library, for a discussion of Coalville, Oakley, and Wanship and the businesses in these towns, including lumber purveyors.
34. Samuel Fortier, "Bulletin No. 38—Preliminary Report on Seepage Water and the Underflow of Rivers," *UAES Bulletins*, Paper 7, http://digitalcommons.usu.edu/uaes_bulletins/7/; Jay D. Stannard, "Irrigation in the Weber Valley," in *Report of Irrigation Investigations in Utah* (Washington, DC: GPO, 1903); "Map of the Weber River Drainage Basin Showing Canals and Irrigated Land," in ibid., 176.
35. "Great Body of Utah Timber," *[Ogden, Utah] Evening Standard*, July 18, 1911.
36. "Park Float," *Park [City, Utah] Record*, October 16, 1909; "Park Float," *Park [City, Utah] Record*, May 25, 1912; "Park Float," *Park [City, Utah] Record*, November 16, 1912; see also "The Life Story of Joseph Hyrum Petersen," MS 12977, Church History Library; "Hundreds of Anglers Swarm Utah Streams," *Salt Lake Herald-Republican*, June 16, 1915; "Speaking of Fish," *Salt Lake Telegram*, July 14, 1931; and "Now a Famed Fisherman," *Park [City, Utah] Record*, July 13, 1934, for fishing accounts in the region in the early 1900s; and "To Extend Trackage," *Salt Lake Herald*, July 24, 1904; and "Park Float," *Park [City, Utah] Record*, August 19, 1905, for camping expeditions along the Weber.
37. "Dynamiters at Work in Weber River," *Deseret Evening News*, July 9, 1910.
38. "Trout Fishing Tuesday," *Salt Lake Telegram*, June 12, 1920.
39. "Irrigation and Reclamation of Arid Lands: Utah Division," August 19, 1889, Sec. 425, in U.S. Congress, *Serial Set* (Washington DC: GPO, 1890), 5, http://books.google.com/books?id=yYY3AQAAIAAJ&pg=RA2-PA5&lpg=RA2PA5&dq=%22all+navigable+rivers,+within+the+Territory+occupied+by+the+public+lands%22+utah&source=bl&ots=kmktpNBOUC&sig=bXu_hco6RDZlHeiXhRoVGYV6_c4&hl=en&sa=X&ei=4mOWUYryD6abiALCooG4Bg&ved=0CDcQ6AEwAg#v=onepage&q=%22all%20navigable%20rivers%2C%20within%20the%20Territory%20occupied%20by%20the%20public%20lands%22%20utah&f=false.
40. *Constitution of the State of Utah*, article XVII, section 1, May 8, 1895, in *Revised Statutes of Utah, 1898* (Lincoln, NE: State Journal Co., Printers, 1897), 37–72.
41. *Adams v. Portage Irrigation Co.*, 72 P.2d 648, 652–53 (Utah, 1937). Several other Utah cases also confirm these principles.

42. *J.J.N.P. Co. v. Utah*, September 22, 1982, http://ut.findacase.com/research/wfrmDocViewer.aspx/xq/fac.19820922_0002.UT.htm/qx.
43. Associate Chief Justice Durrant, "Opinion," July 18, 2008, http://www.utcourts.gov/opinions/supopin/Conatser071808.pdf.
44. Reed Watson, "On the Lookout: Conatster [*sic*] v. Johnson Threatens Stream Bank and Property Rights," *PERC Report* 26, no. 3 (Fall 2008), http://perc.org/articles/conatster-v-johnson-threatens-stream-banks-and-property-rights. For a detailed legal discussion of Utah's stream access laws, see Jeremiah Williamson, "Stream Wars: The Constitutionality of the Utah Public Waters Access Act."
45. Brian Maffly, "Legislature Decides to Bow Out While Courts Deal with Public Access Issues."
46. Derek P. Pullan, "Ruling and Order on Cross Motions for Summary Judgment Re: Plaintiff's Standing and the Public Trust Doctrine," March 8, 2013, http://utahstreamaccess.org/usac-wp/files/PullanMar2013Ruling.pdf.
47. "Trout Fishing Tuesday," *Salt Lake Telegram*, June 12, 1920.

10

The Return of Glen Canyon

The Beginning of a More Sustainable Future for the Colorado

ANNETTE MCGIVNEY

AFTER SEVEN HOURS of hard driving I was tired and perhaps a bit delusional. It was May 2003 and my friend Eli and I were pulling into Hite Marina on the northern end of Lake Powell around 10:00 PM. I had been to the marina several times before in the previous decade but always in broad daylight, when it was bustling with sunburned revelers stumbling off their docked houseboats to get gas and cold drinks. I kept my eyes peeled for these partiers as I drifted in velvety blackness along the marina road looking for a campsite. It was so dark that even the high beams didn't help much.

I squinted into the tunnel of light, continuing down the blacktop. We didn't want to pay to stay at the park campground, so we kept driving toward the lake. But what I thought was the road abruptly became white and broad and looked like rippled concrete.

"Are we driving down the boat ramp?" I asked Eli, not believing my tired eyes.

It was too dark to see what was far ahead or around us. But now I was fully awake, and we were definitely on the boat ramp.

"Keep going!" urged Eli, excited by the prospect that several years of drought had lowered Lake Powell much more than we expected.

At the end of the high beams was brown dirt, not black lake. We kept driving off the end of the boat ramp and onto the lakebed, where boats once docked far above our heads. Certain that the lake had to be nearby and waiting to swallow my car, we parked and set out on foot.

We followed the narrow shafts from our headlamps and leapt from one hardened silt island to the next, across a parched lakebed that had been at the bottom of deep blue waters a few years before. The scene was surreal, and I did not trust my senses. I heard a roaring sound like a wind gust in a forest. Eli heard it too.

We had walked perhaps a quarter mile, and the sound was growing louder. I reasoned that it was some kind of boat generator or machinery at the dock. But where was the dock? Where were the boats? Our headlamps illuminated buoys and old boat anchors stranded on the dry lake floor.

The sound was so loud now that I could barely hear what Eli, who was 100 yards ahead of me, was saying.

"What?" I shouted. He had reached the source. It was something I thought I'd never see or hear in my lifetime.

"It's the Colorado!" he shouted back. "It's flowing!"[1]

Before the floodgates on Glen Canyon Dam closed in 1963, Glen Canyon was a slickrock wonderland unlike anywhere else on the planet. Writer and eco-pundit Ed Abbey succinctly described it as "the living heart of canyon country."[2] Compared to the roiling rapids of Cataract Canyon above and Grand Canyon below, the 180 miles of flat water through Glen Canyon was the wilderness paddling equivalent of curling up on the couch with a pint of Ben and Jerry's. The river moved swiftly but gently between sheer canyon walls 1,000 feet high. Around every bend was another twisting side canyon that harbored its own unique wonderland: alcoves as big as any European cathedral; dark grottos dripping with sweetwater springs and translucent green algae; hanging gardens of maiden hair ferns spilling out of cliff faces; groves of ancient cottonwood trees set against swirling slickrock that reflected and filtered sunlight in so many hues of amber.

"In this dreamlike voyage any unnecessary effort seems foolish. Even vulgar," wrote Abbey in his essay "Down the River" that chronicled a trip through pre-dam Glen Canyon. "Ah, so wild, lonely, sweet, primeval and remote...Every bend in the canyon promised new visionary delights. It seemed to us that Glen Canyon became more beautiful and wonderful with every mile."[3] It wasn't just the scenery that impressed Abbey and others but the size. The massive wilderness straddling the Utah/Arizona border encompassed Glen Canyon itself—as long as the Grand Canyon—as well as hundreds of sinuous side canyons that offered more opportunity for exploration than anywhere else in the Southwest.

The San Juan, Dirty Devil, and Escalante Rivers all flowed into Glen Canyon, creating a unique riparian vortex that pulsed with life. And everywhere ancient Ancestral Pueblo ruins stood as a testament to the way Glen Canyon had nurtured human inhabitants for thousands of years. During the federally authorized "salvage" surveys conducted by the Museum of Northern Arizona and University of Utah during the late 1950s, more than 3,000 archaeological sites were identified in Glen Canyon, mostly granaries, dwellings, and rock art dating between AD 500 and 1250.[4]

Former Sierra Club executive director David Brower famously pronounced Glen Canyon dead after he took his first river trip there just before the floodgates came down. "Glen Canyon died in 1963 and I was partly responsible for its needless death," he wrote in the introduction to *The Place No One Knew*, which was published by the Sierra Club as a postmortem lesson about what was being lost to the insatiable post–World War II development machine. "So were you. Neither you nor I, nor anyone else knew it well enough to insist that at all costs it should endure. When we began to find out it was too late."[5]

Growing up in the 1960s and 1970s when Lake Powell was filling, I always reasoned that the destruction of Glen Canyon was collateral damage in nothing less than the comfortable Southwest lifestyle I enjoyed. After all, it was a modern marvel to have an endless supply of water, air conditioning, and fresh produce in a city like Phoenix that received less than nine inches of rain a year. Plus, I had admittedly enjoyed boating and camping trips on the placid, vast waters of Lake Powell. Yet the tradeoff always seemed bittersweet. I wished that I could have explored the spectacular twisting slot canyons and paddled the lazy river and seen the intact Pueblo ruins, but I was born too late. For me, Glen Canyon was the stuff of legend. I could read about it in the writings of Abbey or John Wesley Powell but would never experience it for myself.[6] At least that is what I believed before the effects of climate change kicked in.

In January 2000 Lake Powell was at 95 percent of capacity with about 24 million acre-feet of stored water and nearly 2,000 miles of shoreline.[7] By April 2005, after six years of significantly below average precipitation in the Upper Colorado River basin, the lake level had dropped a record 140 feet to only 30 percent of capacity.[8] While above-average precipitation in 2005 and 2011 gave Lake Powell storage a slight bump, eight of the last ten years have experienced below-average precipitation in the Upper Colorado River basin. The 2012 water year was 45 percent of average, and in August 2013 the Bureau of Reclamation predicted that another dry year in the Colorado Rockies would produce annual inflow to Lake Powell that was only 40 percent of average.[9] If the forecast for 2013 proved correct, Bureau of Reclamation hydrologists said that the period from 2000 to 2014 would be the driest fourteen-year period on record for the Upper Colorado River basin.[10]

The result of all these dry years is that the second largest reservoir in the United States has remained at half full or less for the better part of a decade. At full pool Lake Powell was 250 square miles and pushed up like octopus tentacles into hundreds of side canyons. But as the reservoir has dropped, hundreds of miles of side canyons on the outer reaches of the reservoir are seeing the light of day for the first time in decades.[11]

After that May 2003 trip to Hite, I embarked on a decade-long expedition to see what had surfaced.[12] It led me to explore not only the canyons that had emerged but also the western water policy that had buried them in the first place.

Could the heart of Abbey's canyon country still be beating, I wondered? And if we are getting a second chance to experience Glen Canyon, could we also be given a second chance to save it?

The water policies behind Glen Canyon Dam date back to a 1922 compact between western states called the Law of the River.[13] As California politicians pushed the federal government to build Hoover Dam, the six other states in the Colorado River basin worried that they were going to lose their ability for future development if California claimed their water rights. Western water law dictates that whoever uses the water first can claim rights to it, even if it's being pumped from hundreds of miles away. So the compact divided Colorado River water evenly between the upper basin (Wyoming, Colorado, New Mexico, Utah) and the lower basin (Arizona, Nevada, and California). The average annual flow of the Colorado was determined to be 15 million acre-feet; the upper basin got 7.5 and the lower got 7.5. The dividing line between the two basins was arbitrarily drawn at Lee's Ferry, Arizona, just above the Grand Canyon.

After the completion of the 726-foot-high Hoover Dam in 1931, congressional delegations from upper basin states also wanted big water storage and irrigation projects to spur growth in the sparsely populated mountain West. The problem for the federal government was that the projects were becoming prohibitively expensive, with little chance that they could ever be paid back by farming operations or municipalities using the water. The Bureau of Reclamation's proposed solution came in 1946 with the introduction of the Colorado River Storage Project and accompanying report titled: "The Colorado: A Natural Menace Becomes a National Resource."[14] Here the concept of building a series of "cash-register dams" was introduced, which would generate revenue for more dams and irrigation projects through the sale of hydropower. Engineers put bull's-eyes on remote sections of the Colorado River that hadn't been practical locations for irrigation or municipal water development but were ideal for hydropower dams because of the high flows and sheer, narrow canyons. Dams and reservoirs were planned for Echo Park in Dinosaur National Monument, as well as Glen Canyon and two locations in Grand Canyon.

While David Brower and the Sierra Club successfully fought to have the dam projects in Dinosaur and Grand Canyon mothballed, Glen Canyon dam moved forward, if only because it was located in a place that Brower and other environmentalists had never been. Brower would later say that his decision not to fight for Glen Canyon until it was too late was the biggest regret of his life.[15] Former Arizona senator Barry Goldwater, one of the staunchest supporters of Colorado River irrigation projects in the 1950s and 1960s, also publicly admitted his misgivings just before his death in 1998. He said on a PBS documentary called *Cadillac Desert: An American Nile* that he regretted his vote supporting Glen Canyon Dam: "If I had it to do all over again, I'd vote against it. Water supply is important but not that important."[16]

At 710 feet high, Glen Canyon was the Bureau of Reclamation's last big dam and marked the end of the agency's empire-building phase. Over the last fifty years the bureau's civil engineers and developers have been replaced by hydrologists and statisticians who methodically manage the West's impounded waters with the assumption that it's all they have to work with. Six dams are located along the Colorado River's 1,400-mile course as well as numerous other diversions and dams on the river's major tributaries. Like a railroad moving freight, the bureau's upper and lower basins develop annual and monthly operating plans that determine exactly how much water on the highly regulated Colorado will move downstream.

The water is released in conveyer belt fashion, with Denver and Salt Lake City drawing their allotment from the Colorado's mountain drainages well above Lake Powell. The thirsty desert meccas of Phoenix, Las Vegas, San Diego, and Los Angeles—along with southern California's massive agricultural irrigation districts—draw their legal allotment from Lake Mead and the reservoirs below. Just above the basin dividing line is Lake Powell, which serves as an insurance policy—a holding tank—to make sure the upper basin has the water that it legally owes the lower basin right there in the bank, allowing Colorado, Utah, New Mexico, and Wyoming to draw freely from water resources upstream. No water is drawn directly from Lake Powell except for the relatively small amount that goes to serve the city of Page (population 7,000) and the cooling towers at the nearby Navajo Generating Station.

But this well-oiled water delivery machine has some kinks. First, the river is overallocated. When the states drew up the Colorado River Compact in 1922, the data used to determine the average annual flow of the river was based on the wettest period (in the early 1900s) in the last 500 years. What was believed at the time to be a conservative estimate of 15 million acre-feet of total annual flow per year is in reality closer to 13 million acre-feet on average during the last five decades. And according to a 1944 treaty an extra 1.5 million acre-feet per year must be delivered to Mexico, which is also in the Colorado River basin but was not included in the 1922 compact. That leaves an annual average deficit of 3.5 million acre-feet.

Currently the lower basin is sucking up 110 percent of its annual 7.5 million acre-feet of Colorado River allocation while the upper basin is only using slightly more than half of its share. The deficit between what is supposed to be and what is has been shouldered completely by the upper basin states.[17] Lately even the role of Powell as a way to bank water reserves has problems. In 2012 the required 8.23 million acre-feet (maf) of water was released from Powell to the lower basin as required by the Law of the River. But due to upper basin snowpack that was 67 percent of average only 4.91 maf flowed into the reservoir.[18] And 2013 was proving to be another record dry year, so Bureau of Reclamation water managers projected only 4.33 maf of inflow, which, after the legal downstream obligations were met, would take Lake Powell down to 43 percent of capacity by October 2013.[19]

"Powell is a buffer against drought. It ensures that the upper basin can meet its obligations to the lower basin," says Bureau of Reclamation hydrologist Katrina Grantz, who manages projections for Lake Powell.[20] When publishing its "Inflow Forecasts and Model Projections," the bureau only predicts one year into the future and relies heavily on the recent past (the last five–ten years) as an indicator for what is to come. "I think what is fairly certain is that there will be increased variability," adds Grantz. "There could be extreme dry years, but also extreme wet years."

Paleoclimatologist Connie Woodhouse, who conducts research for the University of Arizona's CLIMAS (Climate Assessment for the Southwest) program, takes a longer-term view about what might be happening to the West's water supply: "Reservoirs are a good buffer against short term drought."[21] "But our paleoclimatic reconstructions show drought that occurred in the Southwest was far worse than anything we've seen recently. Under those circumstances, our current water infrastructure would be really challenged." Using tree ring data, Woodhouse determined that "persistent drought" lasted from AD 800 to 1300, with the worst being a nonstop sixty-year dry period during the 1100s.[22] And the drought was widespread across all river basins in the West. Could the drought in the 1100s offer a model for a current worst-case scenario? "Maybe," says Woodhouse. But she also notes that it has been hotter over the last decade than the warmest period that she reconstructed from tree ring data. In 2011, for example, the Phoenix area set a record for having more than 100 days over 100 degrees Farenheit. And increased heat further stresses water supply.

While Powell once served as a kind of savings account that allowed the upper basin states to bank excess water, over the last decade it has functioned at best more like a fluid checking account or at worst like an infrastructure that is living off its dwindling savings. When Lake Powell was full, it held four years' worth of annual Colorado River runoff. In August 2013 that amount was down to two years or less. And other problems are associated with maintaining a giant reservoir like Powell in the middle of a desert: a vast amount of the stored water evaporates into the dry air; sediment from the silt-laden Colorado is trapped in Powell and is filling in the upper part of the reservoir; and the environmental effects from the dam are destroying the riparian ecosystem downstream in Grand Canyon.[23] As the levels of Lakes Powell and Lake Mead drop and other issues related to the 1950s-era water infrastructure become more obvious, some environmental groups are calling for change. One of the most outspoken advocates for reform is Richard Ingebretsen, the founder and president of the nonprofit Glen Canyon Institute.

"We need to do what is right for future generations. We can no longer justify having these big reservoirs," says Ingebretsen, who believes that the recovering side canyons in Glen Canyon that have emerged over the last decade should be formally protected.

> Future generations are not going to hold the builders of Glen Canyon dam responsible, because they didn't know any better. But they are going to hold the people who are in charge right now responsible. We have to act to protect Glen Canyon while we still have the chance. Our children and grandchildren are going to look at extinct species and a silted-in Glen Canyon and a dead Grand Canyon, and they're going to ask, "What were we thinking?" History will not be kind.[24]

By the time we made it to a ridge overlooking a broad canyon called Smith Fork, my eight-year-old son, Austin, had experienced his fill of Lake Powell. We were on the fourth day of a five-day boating and hiking expedition in April 2005 when Lake Powell had dropped to its lowest level in decades. I wanted to explore as many reemerged side canyons in Glen as possible. This could be a once in a lifetime event, so I wanted Austin to see it too. But he had grown weary of the beating sun, the boat engine exhaust, and the desert of open water. "Where are we going now?" he asked, skeptical about what we might be doing next on this wild goose chase. He scanned the barren sea of slickrock around us and scowled. Where was the fun? I pointed to the glowing green corridor some 300 feet below that was nestled in the folds of a twisting orange canyon.

We stepped into the drainage right at the lake's old high-water mark. Up-canyon were sinuous narrows and a tunnel-like inlet that harbored a dark swimming hole, but we headed down-canyon to explore the resurrected two-mile section of Smith Fork. It was a world apart from the monotonous barren rock and trapped water of the reservoir. We splashed through a newly flowing spring-fed stream that sparkled in the sunlight. Young cottonwood and willow trees had taken root since the reservoir withdrew and now lined the canyon, along with towering alcoves sheltering fine, cool sand.

"I love this place," Austin proclaimed as he scooped up tadpoles. After just ten minutes of hiking we had landed in a bottomless toy box of natural wonders.

Before the dam, biologists chronicled 500 species of plants and animals in Glen Canyon, including more than 150 different birds.[25] It was an oasis in the desert, a celebration of ecological diversity. From where we sat in Smith Fork on that day in April 2005 it appeared that much of the life had returned, flocking like us to the ribbons of green. We spent the day lazing about in this new eden, watching the parade of life in a place that had been submerged by the reservoir just a few years earlier. Hummingbirds dive-bombed around our ears; a fat beaver slipped into the stream; a bobcat bolted across a sandy terrace; we made toys out of willow twigs and sprawled flat on our backs watching "water music" shimmer its reflected light on the overhang above us. In search of the perfect lunch spot, we stepped through a curtain of cattails and discovered a side drainage of algae-slick sandstone chutes that tumbled into green pools where dragonflies danced.

No one—including bureau engineers, municipal water managers, marina owners, politicians, and mainstream environmentalists—expected ever to see Glen Canyon reemerge alive and well the way it did as Lake Powell withdrew between 2001 and 2005. It was not the ruined, stinky silt flats filled with boat debris that people expected, but a landscape that, at least in the side canyons, resembled the pre-dam riparian splendor famously featured in the Sierra Club's book *The Place No One Knew*. Ironically, a big reason for the healthy ecological recovery was that lake water seeped out of the porous sandstone walls for months after the reservoir had subsided, which jump-started the growth of native vegetation and hastened the return of birds, insects, and animals. Flash floods from winter rains also helped flush out sediment banks in side canyons.

But unlike nearby national parks such as Grand Canyon and Zion, which are managed with the overarching mission to protect the natural resource, the 1.2 million–acre Glen Canyon National Recreation Area is mainly managed to accommodate motorized boating on a fake lake. Even though the more than 100 square miles that have emerged below Powell's highest high water mark represent the size of Arches National Park, the national recreation area mission has no policies to protect the emerging natural environment in Glen Canyon. While the National Park Service, for example, authorized the dredging of the "Castle Rock Cut" near Wahweap Marina in spring 2013 to make it possible for boats to get through as the lake dropped, no Park Service inventories of the ecology or possible exposed archaeological sites in Glen Canyon's reemerged side canyons were taking place. A big reason behind this disparity is the political influence of the motorized boating industry and the strong convictions from the business community of Page that its economic survival depends on motorized recreation in Lake Powell.

Founded in 1956 to house dam construction workers and their families, Page now has about 7,000 residents. Luxury hotels, golf courses, fast food restaurants, gas stations, grocery stores, and a Super Walmart line the manicured streets. According to a study by Michigan State University that was conducted for the National Park Service, visitors to Glen Canyon Recreation Area in 2011 spent $238 million, which supported 2,819 local jobs in the Page area.[26]

In this respect you could still call Glen Canyon a cash-register dam. Motorized recreation on the reservoir has been a big money maker, especially for Glen Canyon National Recreation Area's concessionaire ARAMARK, which runs all marinas, boat rentals, stores, gas stations, lodging, and tours at Lake Powell. ARAMARK is a Fortune 500 company with multimillion-dollar annual profits. It operates concessions at many national parks and other recreation destinations across the country. The company's registered trademark for Lake Powell is "America's Natural Playground."

Compared to the days when Abbey floated through Glen Canyon, boating on Lake Powell clearly makes the "playground" more accessible and full of modern comforts—if you can afford it. A timeshare in a standard houseboat starts at around $20,000 a year. Owning the houseboat outright, or any kind of motorboat big enough to negotiate the wind-whipped waters of Lake Powell, is a major investment because you've also got to pay for boat storage, maintenance, and a vehicle that can haul it. Renting a basic 62-foot-long houseboat for four days costs about $5,500 or $8,300 for a week; a 19-foot motor boat is $400 a day.[27] Then there's the cost of gas—which in March 2013 was $5.44 a gallon at ARAMARK marinas. And gas alone can easily add up to $1,000 per week, because motorboats only get about three miles to the gallon and houseboats get even less.

Compared to Lake Powell's recreation heyday in the 1980s and 1990s, it seems that the recent economic downturn, combined with news about dropping lake levels and what may be changing preferences in outdoor recreation, is taking some of the ring out of the register. Visitation to Glen Canyon National Recreation Area markedly decreased from a high of 3.6 million annual visitors in 1992 to 1.8 million in 2006. In 2013 it was up slightly to 1.9 million visitors. Meanwhile visitation at nearby national parks has remained steady or climbed slightly during the same period. Grand Canyon saw 4.2 million annual visitors in 1992 and 4.5 million in 2013. Zion National Park had 2.3 million visitors in 1992 and 2.8 million in 2013.[28]

"Glen Canyon is the future of this region, and Lake Powell is the past," argues Glen Canyon Institute's Richard Ingebretsen. He maintains that local economies as well as the environment would be better off if the federal government shifted its recreation and water management policies to protect recovering Glen Canyon. Ingebretsen points to nearby Grand Canyon and Zion as evidence that tourists are more interested in visiting the Southwest's crown jewels than in houseboating on a reservoir. "We are presenting a solution to the people of Page, not a threat. Visitors would come from all over the world to see a restored Glen Canyon."[29]

After the hike in Smith Fork with my son Austin on that April 2005 trip, we ventured the next day into a long, twisting drainage called Llewellyn. Austin and I splashed down Llewellyn's recovered stream channel below Powell's old high-water mark. The late afternoon sun cast the orange slickrock under our feet in a fiery glow, and we attempted to identify the bounty of late-spring blooms along the grassy banks: evening primrose, prickly pear, globe mallow. All around us it was wet and green and luscious. Ed Abbey wrote that Glen Canyon was "a portion of the earth's original paradise."[30] And it seemed that we were glimpsing that magic in Llewellyn.

Then we spotted something else. It was big and white and glinting in the day's last sun. Wedged in the sandy bank and almost swallowed by a green tangle of reeds was a twisting fiberglass pool slide. Austin and I walked up and took a closer look. Perhaps just five years ago the slide had been attached to a houseboat floating some fifty feet above where we were hiking. Now it was a relic. It stood like a tombstone marking a forgotten grave.

Since 2007 the Bureau of Reclamation has operated Lake Mead and Lake Powell under an "equalization" policy that still maintains the nearly century-old Law of the River mandates but also implements "shortage guidelines" for sending more water down to Mead if it drops below a certain level.[31] In other words, during a new period of decreasing precipitation and shrinking reservoirs the lower basin needs more assurances, for example, that the taps of residents of Las Vegas won't run dry. As Powell has remained half empty for most of the last decade, so has Mead, which is the nation's largest human-made reservoir, able to hold up to 28 million acre-feet of water.

In June 2010 Lake Mead dropped to 39 percent of capacity, the lowest since the reservoir started filling in 1937. It was just thirty-two feet above the water intake for metropolitan Las Vegas. If the reservoir had dropped another seven feet, the bureau's shortage guidelines would have kicked in: water supplies to Phoenix, Tucson, and Las Vegas would have been cut by nearly a third. But a wet winter in the Colorado Upper Basin helped avert a potential crisis, and releases from Lake Powell in 2010 and 2011 "equalized" the level in Mead. In August 2013, after another dry winter, the bureau projected that total water storage in the Colorado River basin would end the year (in September) at 49 percent of capacity. The back-to-back record dry years also caused Bureau of Reclamation water managers to predict for the first time in the agency's history that there was a 50 percent chance that water deliveries to the lower basin would be cut in 2016.[32] In other words, the two years' savings account in Powell would be spent down. According to the Law of the River, Nevada and Arizona are first and second on the chopping block if the bureau's shortage guidelines kick in. Las Vegas relies almost completely on the water supply from Lake Mead. The Phoenix and Tucson metro areas depend on Mead as well as on reservoirs from the Salt and Verde Rivers, which had also dropped to 55 percent capacity in August 2013.[33]

Whether the dropping reservoir levels are due to extended drought or a new climate regime caused by global warming, the potential implications of not having adequate water in the West are huge. Nearly 40 million people rely on the Colorado River for municipal water supply, and 5.5 million acres of farmland are irrigated with its runoff. In a December 2012 report the Bureau of Reclamation acknowledged the potential threat: "Looking ahead, concerns regarding the reliability of the Colorado River system to meet future Basin resource needs are even more apparent, given the likelihood of increasing demand for water throughout the Basin coupled with projections of reduced supply due to climate change." The report projected that Colorado River flow could decrease by 9 percent in the next fifty years while population in the basin would likely increase, producing a potential water supply deficit of 3.2 million acre-feet by 2060.[34]

Other recent reports on water supply and climate change in the West are more dire. A 2012 report based on a study by scientists from Princeton, Colorado State University, and the U.S. Forest Service found that Lake Powell and Lake Mead could dry up completely by 2080 if consumption increased with projected population growth while precipitation decreased.[35] A 2008 report from Scripps Institution of Oceanography scientists projected a 50 percent chance that Lake Mead could go dry by 2021.[36] And a 2009 study predicted that climate change could reduce Colorado River flow by 20 percent and cause the basin's reservoirs to be fully depleted by 2057.[37]

"I am not all gloom and doom. The Colorado River itself is not going to go completely dry," says Zach Guido, a scientist at University of Arizona's CLIMAS.[38] "There will be water in the river, but management is going to have to adapt to less in reserve. Who will the winners and losers be under these circumstances? We can't just cater to the winners. We need to move away from the big infrastructure, be more nimble and get creative." The biggest traditional winner in the western water equation is agriculture, which claims nearly 90 percent of the Colorado's upper basin allotment and 80 percent of the lower basin's.[39]

A solution proposed by the Glen Canyon Institute to move away from the West's current clunky water infrastructure is simply to "fill Mead first."[40] Rather than having two giant reservoirs that are half empty or less, the group says that it would be most efficient and environmentally sound to keep more water in Mead and less in Powell. A recent study conducted for the institute shows that keeping Powell at an elevation of 3,490 feet above sea level (full pool is 3,700 feet) would save or recover up to 300,000 acre-feet of water a year that currently seeps into the reservoir's sponge-like Navajo sandstone banks.[41] The igneous rock around Mead is not nearly as porous. Glen Canyon Institute's projections on potential savings from reduced seepage in Powell reflect the net total after factoring in surface area evaporation rates for both reservoirs. Currently water lost to evaporation in Powell and Mead combined is 1.6–1.8 million acre-feet per year. Under the Fill Mead First Plan, evaporative loss in Powell would be reduced due to a smaller surface area but slightly increased in Mead with a full lake level in a warmer desert climate. Even when factoring in higher evaporation rates in Mead, however, the institute estimates that total water lost to evaporation in both reservoirs would be reduced to 1.3 million acre-feet a year if Powell was kept at about 25 percent capacity.[42]

"The fill Mead first plan would drop Powell to 65 feet below what it was at its lowest level in 2005, so a large amount of the side canyons in Glen Canyon would be out of the water," explains Glen Canyon Institute program director Michael Kellett. "And it would also stabilize the water level so that it would no longer allow the rising and lowering reservoir simply for water storage convenience. The main river channel would still be inundated but most of the side canyons would be permanently out of the water and allowed to recover."[43] At this proposed reservoir elevation, Kellett says that hydropower could still be generated and the reservoir would store up to 6.1 million acre-feet of water and occupy a surface area of 52,600 acres—which is 20 percent larger than Flaming Gorge, the second largest reservoir in the Colorado River upper basin. Plus, according to the institute's data, the amount of water saved or recovered from reduced seepage in Powell would add as much to the system as Nevada's entire current annual Colorado River allotment.

After completing its "Colorado River Basin Water Supply and Demand Study" in 2012, the Bureau of Reclamation is formulating its "Next Steps" phase. "We had a lot of stakeholders sit around the same table for the first time to complete the study," says the bureau's Carly Jerla, who was co-study manager of the multiyear project.[44] "We have traction now but we have to keep going. We have to find ways to reduce the uncertainties for the future." In March 2013 the bureau was planning to hold more meetings with Colorado River stakeholders and was looking at first implementing what Jerla called "no regrets" steps involving municipal and agricultural conservation. But the proposal to fill Lake Mead first was not included in the agency's complex matrix of proposed solutions.

Dan Beard, a former commissioner of the Bureau of Reclamation who served under President Bill Clinton and is now an advisory board member of Glen Canyon Institute, puts the situation in blunter terms. "The lake is lower now than anybody ever imagined," notes Beard. "It provides a new reality about what is possible. It's time for the federal government to consider operating the dam in a different way so that we can protect what's been uncovered."[45]

Granted, filling Mead first would not completely solve the complex water supply problems that threaten the West. But it would be a significant step in the right direction on both a practical and symbolic level. Like a two-car family that decides to trade in the gas-guzzling SUV and save money by owning just one vehicle, a policy change by the Bureau of Reclamation to maximize storage in Mead offers advantages in terms of economy. But, perhaps even more importantly, it would signify a paradigm shift. Such a move, enacted by government and supported by citizens, would say: we are down-sizing our infrastructure and embracing the reality that we must learn to live with less. And getting Glen Canyon back is icing on the cake.

Hite Marina was high and dry and still closed to boats for another summer season in 2014. But the Colorado River continued to cut its way through a deep silt channel past the marooned boat docks. In side canyons downstream from Hite, cottonwood and willow saplings reached toward the sun. Even Cathedral in the Desert, the most hallowed spot in predam Glen Canyon, emerged from the reservoir. In March 2014, when Lake Powell had dropped to 38 percent capacity, I returned to Glen Canyon and stood on the floor of Cathedral in the Desert, where an island of earth had just surfaced and a 20-foot tall waterfall, once completely buried by the lake, tumbled into a turquoise pool.

While global warming presents some serious predicaments for water managers as well as the 40 million people in the West who rely on the Colorado River for municipal and agricultural uses, the problem of a half-empty reservoir also harbors hope. It offers the promise of recovering the "Place No One Knew" and the potential that comes with living in a way that supports a sustainable cycle of life. Ed Abbey's "heart" of Southwest canyon country is indeed still beating. It pulses through every new bud, every returning song bird, every resurrected stream and resuscitated spring. Here, in the twisting sandstone narrows filled with frogs and ferns and sweetwater seeps, a new era blooms.

Notes

1. The preceding passages are quoted from *Resurrection: Glen Canyon and a New Vision for the American West* (Seattle: Braided River, 2009) and are reprinted here with permission of the publisher.
2. Edward Abbey, *Beyond the Wall: Essays from the Outside*, 95.
3. Edward Abbey, *Desert Solitaire: A Season in the Wilderness*, 177.
4. William Y. Adams, "Ninety Years of Glen Canyon Archaeology 1869–1959: A Brief Historical Sketch and Bibliography of Archaeological Investigations from J. W. Powell to the Glen Canyon Project." See also Don D. Fowler, James H. Gunnerson, and Jesse D. Jennings, "The Glen Canyon Archeological Survey," *Anthropological Papers* 39, University of Utah, Department of Anthropology (May 1959).
5. Eliot Porter, *The Place No One Knew: Glen Canyon on the Colorado* (1963 ed.), 5.
6. John Wesley Powell, *The Exploration of the Colorado River and Its Canyons*, 232–33.
7. As indicated elsewhere in this collection, an acre-foot of water equals 326,000 gallons (enough water to flood an acre of land one foot deep or support 1,200 Utahns for a day).
8. Lake Powell Water Database, http://lakepowell.water-data.com.
9. U.S. Bureau of Reclamation Upper Colorado River Region, "Glen Canyon Dam/Lake Powell, Current Status," http://www.usbr.gov/uc/water/crsp/cs/gcd.html. See also University of Arizona Climate Assessment for the Southwest, "2012 Water Year in Review," *Southwest Climate Outlook* 11, no. 10 (October 2012). See also University of Arizona Climate Assessment for the Southwest, "August Climate Summary," *Southwest Climate Outlook* 12, no. 8 (August 2013).
10. U.S. Bureau of Reclamation Upper Colorado River Region, "Glen Canyon Dam/Lake Powell, Current Status," http://www.usbr.gov/uc/water/crsp/cs/gcd.html.
11. Daniel Glick, "A Dry Red Season: Drought Drains Lake Powell, Uncovering the Glory of Glen Canyon."
12. See my book *Resurrection: Glen Canyon and a New Vision for the American West.*
13. U.S. Bureau of Reclamation Lower Colorado Region, "Law of the River," http://www.usbr.gov/lc/region/g1000/lawofrvr.html.
14. Russell Martin, *A Story That Stands Like a Dam: Glen Canyon and the Struggle for the Soul of the West*, 43–74.
15. David Brower, *For Earth's Sake: The Life and Times of David Brower*, 351.
16. KCET Los Angeles, PBS, and Columbia TriStar Television, *Cadillac Desert: An American Nile, Water and the Transformation of Nature* (Program 2), 1997.

17. U.S. Bureau of Reclamation, "Colorado River Basin Water Supply and Demand Study: Technical Report C; Water Demand Assessment," http://www.usbr.gov/lc/region/programs/crbstudy/report1.html.
18. Upper Colorado Basin Snowpack database, http://snowpack.water-data.com/uppercolorado/index.php?getall=1.
19. U.S. Bureau of Reclamation Upper Colorado River Region, "Glen Canyon Dam/Lake Powell, Current Status," http://www.usbr.gov/uc/water/crsp/cs/gcd.html.
20. Katrina Grantz, hydraulic engineer, U.S. Bureau of Reclamation Upper Colorado Region, telephone interview, March 2013.
21. Connie Woodhouse, professor, University of Arizona School of Geography and Development, telephone interview, March 20, 2013.
22. Connie Woodhouse and Jeffrey Lukas, "Multi-Century Tree-Ring Reconstructions of Colorado Streamflow for Water Resource Planning."
23. Steven P. Gloss, Jeffrey Lovich, and Theodore Melis, eds., *The State of the Colorado River Ecosystem in Grand Canyon: A Report of the Grand Canyon Monitoring and Research Center 1991–2004*, 207–19.
24. Richard Ingebretsen, president, Glen Canyon Institute, telephone interview, March 11, 2013.
25. Gwendolyn Waring, *River and Dam Management: A Review of the Bureau of Reclamation's Glen Canyon Environmental Studies.*
26. Yue Cui, Ed Mahoney, and Teresa Herbowicz, eds. *Economic Benefits to Local Communities From National Park Visitation, 2011*, 19, http://www.nature.nps.gov/socialscience/docs/NPSSystemEstimates2011.pdf.
27. ARAMARK Lake Powell Resorts and Marinas, www.lakepowell.com.
28. U.S. National Park Service, Public Use Statistics Office, "NPS Visitation Database," https://irma.nps.gov/Stats/Reports/Park.
29. Ingebretsen interview.
30. Abbey, *Desert Solitaire*, 174.
31. U.S. Bureau of Reclamation Lower Colorado Region, "Colorado River Interim Guidelines for Lower Basin Shortages and Coordinated Operations for Lake Powell and Lake Mead," 2007, http://www.usbr.gov/lc/region/programs/strategies.html.
32. U.S. Bureau of Reclamation Upper Colorado River Region, "Colorado River Basin Water Supply and Demand Study."
33. University of Arizona Climate Assessment for the Southwest, "August Climate Summary."
34. U.S. Department of Interior Bureau of Reclamation, "Colorado River Basin Water Supply and Demand Study: Executive Summary," December 2012, http://www.usbr.gov/lc/region/programs/crbstudy/report1.html.

35. Romano Foti, Jorge A. Ramirez, and Thomas C. Brown, *Vulnerability of U.S. Water Supply to Shortage: A Technical Document Supporting the Forest Service 2010 RPA Assessment* (Fort Collins: U.S. Department of Agriculture, Forest Service, Rocky Mountain Research Station, 2012), 129–31.
36. Tim Barnett and David Pierce, "When Will Lake Mead Go Dry?"
37. Balaji Rajagopalan, Kenneth Nowak, James Prairie, et al., "Water Supply Risk on the Colorado River: Can Management Mitigate?"
38. Zach Guido, staff scientist, University of Arizona Climate Assessment for the Southwest, telephone interview, March 15, 2013.
39. U.S. Bureau of Reclamation, "Colorado River Basin Water Supply and Demand Study."
40. "Fill Mead First," Glen Canyon Institute, http://www.glencanyon.org/glen_canyon/fill-mead-first.
41. Tom Myers, "Loss Rates from Lake Powell and Their Impact on Management of the Colorado River."
42. Ibid.
43. Michael Kellett, program director, Glen Canyon Institute. email correspondence, March 12 and March 24, 2013.
44. Carly Jerla, U.S. Bureau of Reclamation, manager for Colorado River Basin Supply and Demand Study, telephone interview, March 22, 2013.
45. Beard quoted in McGivney, *Resurrection*, 29.

11

≈

Land of 20,000 Wells

Impacts on Water from Oil and Gas Development in Eastern Utah

HAL CRIMMEL

Twenty years ago I rattled down a rough and dusty road to Sand Wash, site of an old ferry crossing on the Green River in eastern Utah. Located south of the town of Myton, Sand Wash is the place where river trips launch for the 84-mile float down Desolation and Gray Canyons, two remote, spectacular canyons on the Green River. The canyons are still there, but much has changed over the last two decades in the surrounding region. The era of building massive dams such as Glen Canyon and Flaming Gorge is over—for the time being, at least—so the change comes not from finding the canyons underwater or discovering that sprawling new cities fed on cheap water and power have sprung up to the east and north. Rather, it's the intensity of the latest phase of oil and gas drilling that has created and will continue to create the changes. As I've written elsewhere, one of the more significant threats to public lands in and around eastern Utah's Uinta Basin comes from hydrocarbon drilling and production.[1] One specific aspect of those threats concerns the impacts on water resources.

Someone coming from the towering peaks and lush canyons of the Wasatch Front or the otherworldly spires, arches, and narrow canyons of southern Utah's redrock country can be forgiven for not thinking much of the Uinta Basin, with its alkaline soils, sparse vegetation, and bitter cold winters. It is a harsh and unforgiving place. Yet the extreme weather, the astringent ruggedness of the landscape, and its off-the-radar vibe are reasons why I am drawn to the region. In my estimation the Uinta Basin and the lands immediately to the south contain great beauty, with deep canyons, streams, and dramatic erosional features. But that beauty doesn't always immediately conform to conventional notions of scenery.

And the Uinta Basin and the surrounding region have a relatively new form of scenery—industrial scenery, that is—that brings joy to many and anguish to others: thousands of oil and gas wells, pipelines, roads, condensate tanks, and pumping stations. Today, two decades after the date of that first Desolation Canyon trip, I'm headed across that same road strewn with the same tire-puncturing rocks toward Sand Wash. I'm ready to run the river again, all the way down to the takeout located about a dozen miles north of the town of Green River, Utah, where I-70 crosses the river itself.

Before turning off U.S. 40 at the town of Myton, though, I'm surprised by the number of westbound tandem tanker rigs carrying crude oil to the Salt Lake refineries—sixty-five trucks in two hours or about one truck every two minutes. Barring some sudden plunge in crude oil prices, this heavy traffic is likely to continue for a long time. Newfield Exploration Company, for instance, estimates its Uinta Basin assets at over "700 million barrels of oil equivalent." Waxy Uinta Basin crude, which requires special handling and intensive refining, goes for about 80 percent of the price of a barrel of benchmark West Texas Intermediate crude.[2] With current prices around $80 per barrel, 700 million barrels represent a lot of money. In addition, the U.S. Geological Survey estimates that Utah has 400 *billion* barrels of oil shale,[3] with most of it in eastern Utah.[4] And that's not even counting the oil contained in the state's tar sands deposits, estimated at another 12 to 19 billion barrels.[5] Tar sands are typically strip-mined and then require immense quantities of water for processing (anywhere from a 5:1 to 3:1 ratio of water to oil) before the refining process can begin. The Utah Division of Oil, Gas and Mining claims that a recently approved project for the PR Springs tar sands mine in Utah uses only 1.5–2 barrels of water for each barrel of oil produced.[6]

If tar sands mining operations should become widespread,[7] impacts on limited water resources and on local ecosystems in the tar sands regions will be considerable. For instance, in addition to multiple new mines, new roads will need to be established, which fragments wildlife habitat. Truck traffic on these roads releases clouds of dust into the air, impacting air quality. Groundwater supplies are limited, and a drying Colorado River is already overallocated. All this does not include the longer-term effects

Figure 11.1. A network of roads that lead to oil and gas wells continues to spread across the Uinta Basin. This view looks north toward the community of Duchesne and farther toward the Uinta Mountains. Photo by Dan Miller.

on Utah's climate and water resources of pumping even more greenhouse gases into the atmosphere.[8] Estimates vary, but source-to-pump greenhouse gas emissions (GHG) from oil shale operations in Utah (the total amount of GHG from mining, upgrading, transport, and refining of oil shale) are estimated to be 300–400 percent more than GHG emissions from fuels produced from conventional crude oil.[9]

Today much of the crude being hauled in the tanker trucks is coming out of the oil fields south of Myton, a moonscape dominated by nodding pump jacks, above-ground pipelines, trucks billowing clouds of dust, condensate tanks, and other types of structures. I mention the scale and scope of these operations to emphasize that we're not talking about a few scattered oil rigs—these days eastern Utah is a major hydrocarbon player.

The majority of drilling activity takes place in the Uinta Basin, an area encompassing five Utah counties and about 5.8 million acres. For the foreseeable future about 1,000 new wells are scheduled to be drilled in Utah each year.[10] In August 2011 the Greater Uinta Basin already had a total of 15,701 existing wells, with 13,126 listed as active.[11]

In 2011, 28,417 new wells were pending,[12] so the pressure on water resources (and air quality) is expected to triple in coming years. As Ken Kreckel (a former Marathon Oil geologist) notes, the Rocky Mountain region, which includes Utah, already often has 128 wells per square mile, a density unmatched anywhere in the world.[13]

Clearly, by arriving here in a motor vehicle, I am contributing to the impacts of petroleum production and consumption on air and water quality in Utah. Most Utahns are consumers of hydrocarbon-based products, so all share some responsibility for the thousands of wells found in the state. Equally responsible, though not perhaps directly, is anyone with even $100 in a pension fund, retirement plan, or stock mutual fund. About 11 percent of the holdings in a broad-based stock market index fund typically are from the energy sector, which includes oil and gas companies, oilfield service providers, pipeline companies, and the like, all of which do business in the region. This means that even those most fiercely committed to conservation of public lands and water resources may unwittingly be providing capital that funds the operation of the companies responsible for the fragmentation of wildlife habitat, road building in undisturbed areas, light pollution from drilling rigs and trucks, air pollution, and water pollution. I mention these issues not as a form of condemnation but rather to acknowledge the reality of the situation in which many people find themselves. Perhaps concerned about the health effects stemming from hydrocarbon exploration and production, they are also concerned about the effects of these activities on local, regional, and global environments. But they are not sure what, if anything, can be done to manage these effects better. Making a personal choice to limit vehicle or air travel can be a start, though broader initiatives may be the best long-term strategy to protect water resources in eastern Utah; I suggest several ideas here.

But first it might be helpful to identify the short-term and long-term threats that oil and gas operations pose to a fundamental right that most Utahns have taken for granted: clean water in aquifers, groundwater, lakes, reservoirs, streams, and rivers. A citizen of this state who goes fishing or boating—as do millions of Americans across the nation—expects to find the water free of contaminants. Having clean drinking is something most take for granted. No one wants water flowing from a faucet to be contaminated with natural gas or with even the tiniest amounts of drilling or fracking fluids.[14] Nobody wants groundwater or surface water contaminated by "flowback" or "produced water," a waste product that poses one of the greatest challenges to eastern Utah's water resources. According to the United States Department of Energy, produced water

> is water trapped in underground formations that is brought to the surface along with oil or gas. Because the water has been in contact with the hydrocarbon-bearing formation for centuries, it contains some of the chemical characteristics of the formation and the hydrocarbon itself. It may include water from the reservoir, water injected into the formation, and any chemicals added during the production and treatment processes.[15]

Two issues are associated with produced water: quantity and quality. Produced water typically can be highly saline, containing oil and grease as well as "various natural inorganic and organic compounds or chemical additives used in drilling and operating the well, and naturally occurring radioactive material."[16] Saline aquifers far beneath the surface that are not perforated by drilling have little risk of contaminating surface waters such as rivers or shallow groundwater aquifers that supply drinking water. Nor do the toxic compounds or radioactivity present deep underground in oil and gas formations present a risk—that is, until drilling or fracking brings them to the surface as produced water. This produced water must then be disposed of, either by injecting it into wells deep underground or by evaporating it out of above-ground pits.

Widespread energy development can impact water locally but also on a broader scale, crossing state lines. As of 2013, for example, oil and gas producers in Wyoming announced proposals to drill more than 23,000 wells, with 8,000 of those new wells in Wyoming's Green River Basin alone,[17] a region that lies upstream from Utah. Any sort of accidental spill, illegal dumping of produced water, or leaks of crude oil, natural gas, chemicals, produced water, or drilling or fracking fluid in the Wyoming portion of the Green River watershed has the potential to impact those downstream—in Utah. The same is true for any form of water contamination occurring in the oil and gas fields of western Colorado. In turn, mismanagement of oilfield water in Utah will not just impact local ecosystems, such as those found in Desolation and Gray Canyons or the popular boating destination of Lake Powell.[18] Mismanagement, mistakes, or mechanical malfunctions also have the potential to harm drinking water supplies for cities downstream in other states, such as Las Vegas, Phoenix, Tucson, and Los Angeles. The water resources in a watershed—whether surface or groundwater—might seem to be inherently local but in fact often are broadly regional and should continue to be treated as such. All stakeholders should have a say in the decision-making process. Yet regulations are typically managed at the state level, though in the summer of 2013 the Bureau of Land Management circulated a draft proposal of federally mandated fracking regulations.

Before specifically addressing the water-related issues in Utah, let's take a brief look at the process of oil and gas drilling and production. Each new well drilled requires drilling fluid, which functions to bring rock cuttings to the surface, "lubricate and cool the drilling bit, stabilize the wellbore (preventing cave-in) and control downhole fluid pressure."[19] If hydraulic fracking (HF) is used, the well must not only be drilled but then also pressurized with fresh or saline water in order to extract the hydrocarbons. Each time a well is fracked additional water and fracking fluid is needed. As Vikram Rao, a former senior vice-president of Halliburton Corporation notes, for fracking operations "on average about four million gallons of fresh water are used per well, but the use of six million gallons is not unusual."[20] The Utah Division of Oil, Gas and Mining estimates the figure for Utah wells at closer to 300,000–500,000 gallons per frack per well because the lengths of the well bores typically are shorter than in other parts of the country and consequently do not require as much water to pressurize them.[21]

Still, fracking processes use significant quantities of water even in Utah. Also of concern are the amounts and types of chemicals mixed with the water. Industry-sponsored websites such as energyindepth.org and "The Real Promised Land" emphasize that only 0.5 percent of the total volume of substances that go down the well bore consists of chemicals used in drilling or fracking, with the other 99.5 percent consisting of water and sand. Of course, 0.5 percent of 4 million gallons is 20,000 gallons of chemicals for each well drilled or each frack. Even when the numbers are calculated using the Utah figure of 500,000 gallons of water for each frack 0.5 percent of that is still 2,500 gallons of chemicals. Multiply those individual well data by the thousands of wells being drilled and fracked across the Uinta Basin and it becomes clear that very large volumes of chemicals are being introduced into the environment.

A 2011 U.S. congressional report noted that between 2005 and 2009 the major fracking companies used "over 2500 hydraulic fracturing products containing 750 compounds. More than 650 of these products contained

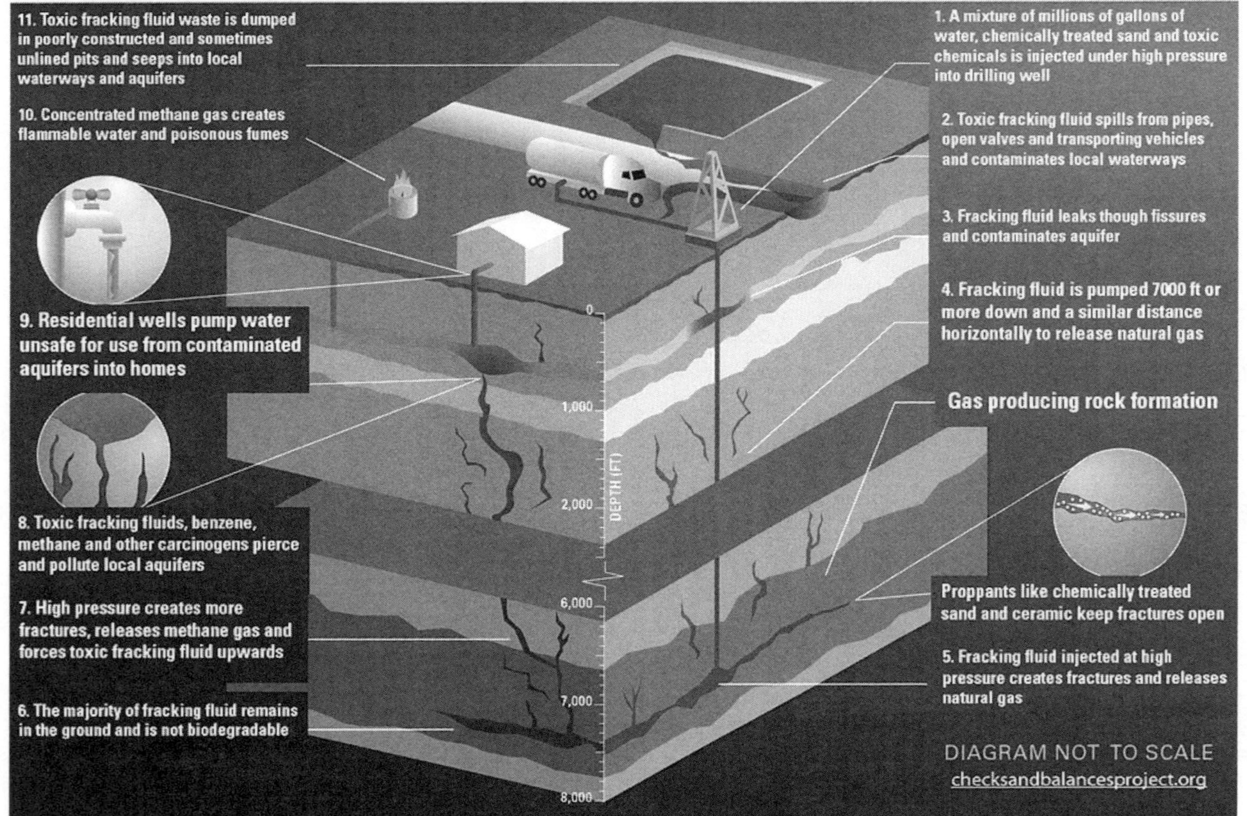

Figure 11.2. Fracking diagram. Checks and Balances Project.

chemicals that are known or possible carcinogens, are regulated under the Safe Drinking Water Act for their risks to human health or are listed as hazardous air pollutants under the Clean Air Act."[22]

In an attempt to make these chemicals seem nonthreatening, some industry sources portray them as "household" chemicals.[23] But these chemicals include ethylene glycol, the toxic main ingredient in antifreeze, along with acids, antibacterial agents, and corrosion inhibitors as well as substances such as methanol, diesel fuel, and other petroleum distillates,[24] among other compounds.[25] I might use such substances in the garage. But surely no one wants to find traces of antifreeze or diesel fuel in drinking water, in irrigation water, or in the rivers and streams where people fish and swim. People suggesting anything to the contrary, as my father used to say, need to have their heads examined.

As I bounce along the dusty road toward Sand Wash, heavy trucks haul equipment through the oilfields. Clouds of dust drift across the countless well sites. With June temperatures near 95 degrees, I am glad to have ethylene glycol in the radiator, but I can't imagine that it would be welcome in a glass of water, even on this hot June day.

Would those who claim that the mix of chemicals used in drilling and/or fracking is safe, such as Colorado governor John Hickenlooper, be willing to drink, cook, shower, and brush their teeth with this water for an entire year? For five years? For ten years? Would they be comfortable drinking the water from Pavillion, Wyoming, where fracking is suspected of contaminating residents' wells? Residents there began complaining about odors in their drinking water after EnCana Corporation started fracking operations. Well testing done by an EPA contractor revealed elevated levels of substances such as gasoline range organics, xylene, toluene, and methane, which were 15 times, 32 times, 35 times, and 355 times higher than federal reporting limits.[26] As indicated, such compounds are present within hydrocarbon formations deep underground, although their sudden appearance in groundwater subsequent to fracking operations would seem to suggest a direct link. But documenting such links conclusively–in other words, legally–is difficult. In the case of Pavillion, reports say that "EPA efforts to find potential pathways from deeper areas where gas is extracted to shallower areas tapped by domestic water wells have been inconclusive,"[27] though it seems likely that the methane, for instance, migrated from deep underground into groundwater.

Such cases help foster repeated claims that no *documented* cases of fracking contaminating groundwater exist. These denials seems to obscure the point. According to Lisa Jackson, the former head of the EPA, there are in fact many cases of "groundwater and drinking water contamination" resulting from drilling operations in general.[28] Though drilling technically is different from the actual process of fracking, it's the entire package of well operations that poses the risks to water resources. Indeed, as Utah's director of the Division of Oil, Gas, and Mining noted in a recent talk, the most pressing issues associated with fracking are not related to the actual frack itself but to the drilling, casing, and cementing of the well and the disposal of fluids.[29] Essential to protecting water supplies is "maintaining the mechanical integrity of the well."[30] Should an active well fail now or in the future, leaks can lead to contamination.

Even abandoned wells pose a risk. "It's uncertain how long oil well casings and plugs will last" and a "recent U.S. Geological Survey study of decades-old wells in Montana found plumes of salt water migrating into aquifers and private wells, rendering the water from them unfit for drinking."[31] Anyone with even a passing acquaintance with concrete knows that over time it can crack and crumble. In seismically active Utah an earthquake of moderate proportions would damage new and older wells alike. Even conscientious operators employing highly trained geologists, engineers, and rig operators following best industry practices can make mistakes, have material failures, suffer blowouts, or not fully understand the relationship between aquifers. In the North Dakota oil and gas fields, for instance, "the depth of the shale formations and the intervening rock layers make it unlikely fracking fluids will migrate upwards far enough to contaminate shallow aquifers. But no one knows for sure,"[32] because fracking can impact different geologic formations in different ways. Caution is warranted, even in conventional operations that do not require fracking or that simply transport crude oil.[33] As a former Halliburton vice-president asks, "Do producing gas wells sometimes leak into freshwater aquifers? The answer is yes."[34] Advanced technology and quality construction are no guarantee of safety, as the Utah director of the Division of Oil, Gas, and Mining notes. "Casing is supposed to last the life of the well. Every now and then you have leaks that need to be repaired."[35] This matter-of-fact statement from a senior government administrator seems to lay it out quite clearly: the process of producing oil and gas is not perfect. The question is what to do about it.

I think that answering this question is made harder by companies and industry trade groups (again, see energyindepth.org or "The Real Promised Land") that utilize the "trust us, it's perfectly safe and always has been" argument. This approach is similar to the approach historically employed by tobacco companies: sowing doubt and "creating controversy and contradiction" over potential health and environmental effects.[36] In 1976, for instance, Philip Morris claimed that "[n]one of the things which have been found in tobacco smoke are at concentrations which can be considered harmful. Anything can be considered harmful. Apple sauce is harmful if you get too much of it."[37] Energy interests have responded to Josh Fox's 2010 documentary *Gasland* with a pro-fracking documentary, *FrackNation*, and the partisan website entitled "The Real Promised Land." Neither offers much in the way of scientifically based evidence, but both do an excellent job of using the tobacco industry's strategy of sowing controversy, confusion, and doubt and attacking those who express concerns about fracking.[38]

Both tobacco use and oil and gas exploration and production are public health issues, so it can be helpful to compare the two. Though there are laws governing tobacco use that protect public health, fracking is exempt from many such federal laws. The Energy Policy Act of 2005, for instance, exempted hydraulic fracking from elements of the Safe Drinking Water Act, the Clean Air Act, and the Clean Water Act. It also exempted companies from having to disclose the chemicals used in fracking operations, which are normally required under the Clean Water Act. We might reasonably ask: if the chemicals and process are absolutely safe, why do fracking operations need to be exempt from the laws that protect the health of humans and the environment?

One bright spot in Utah is that producers are now required by the Utah Division of Oil, Gas, and Mining to list the chemicals used to frack wells, on the Frac Focus Chemical Disclosure Registry, an Oklahoma City–based organization managed by the Interstate Oil and Gas Compact Commission and the Groundwater Protection Council.[39] Listing the chemicals used in each Utah fracking operation is a positive step. But a recent report from Harvard criticizes this registry, claiming that it is full of loopholes that

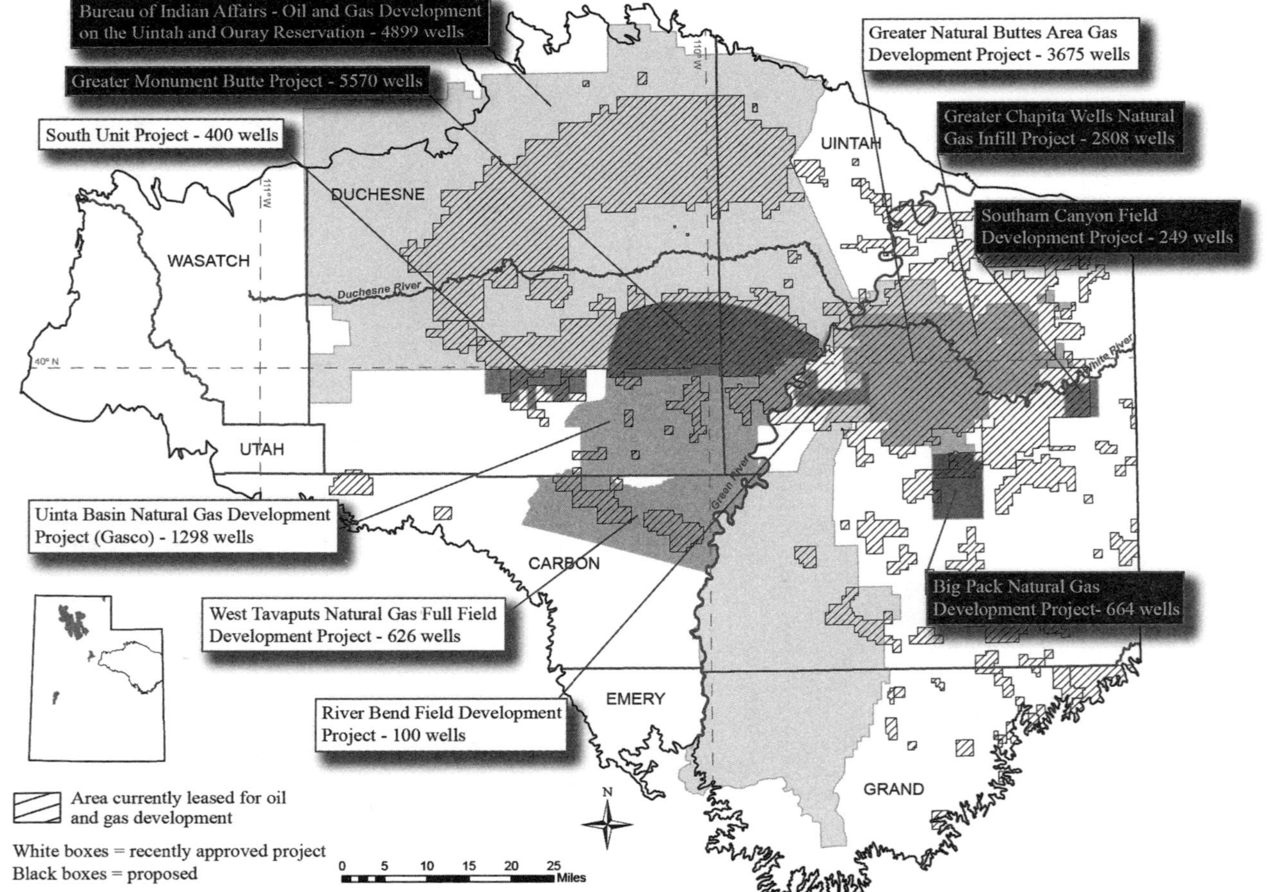

Figure 11.3. Proposed Uinta Basin wells. Utah Geological Survey.

allow operators so inclined to skirt their responsibilities.[40] I would suggest that Utah enact and enforce its own water-related regulations that exceed federal standards and require producers to adhere to them. No one is going to walk away from Utah's multibillion-dollar hydrocarbon bonanza to do business in a "more business-friendly state," leaving millions of barrels of oil in the ground and billions of dollars on the table.

So oil and gas drilling and fracking procedures use water that is already limited in eastern Utah and also use volumes of chemicals that no one would want to find in surface or groundwater used by humans and wildlife. But perhaps a bigger problem has to do with the amount of water that must be either disposed of or recycled from each well.

Two of the top three concerns of the Utah Division of Oil, Gas and Mining related to water have to do with hydrocarbon carryover and saline water surface disposal.[41] Hydrocarbon carryover results when hydrocarbons are not completely separated from produced water, leaving that water polluted and complicating its disposal. Producers typically separate oil and gas from water on site, but the cleaning process is not always perfect. Often that leftover water is dumped into evaporation ponds, which can leak or be damaged by flash floods. As the water evaporates significant amounts of volatile organic compounds (VOCs) such as benzene can be released into the air, worsening regional air pollution. Saline water, which underlies much of eastern Utah, comes up the well bore during oil and gas operations as produced water. This water must be disposed of, typically in deep wells. But Utah's "current disposal wells are at or near capacity," according to Utah Geological Survey geologist Michael D. Vanden Berg,[42] and 94 percent of produced saline water is injected into these wells.[43] Trying to find suitable geological strata and the appropriate depth to drill new disposal wells that will not allow saline water to contaminate groundwater is a major concern. One issue, for instance, is the cross-cutting of gilsonite veins, which can conduct saline water into freshwater aquifers.[44] Even with sophisticated mapping technologies that identify rock formations, extensive core samples, and so forth, it's often difficult to conclude with certainty which formations are leak-proof and will keep saline water contained and which will not.

Some companies are recycling their produced water, which eliminates the need to inject it into deep wells for disposal. Andarko Petroleum and

Newfield Exploration, which recycles 10,000 barrels of produced water per day at its Sand Wash facility, are two of the companies operating in the Uinta Basin that have begun to focus on recycling produced water. Such companies seem to embrace forward thinking that considers environmental concerns as part of the new triple bottom line: people, profit, planet. At least two companies operating in Utah are recycling produced water, so why not make recycling produced water the industry standard? The state could require all companies operating in Utah to recycle their produced water, thereby reducing or eliminating the need for injection wells to dispose of produced water. Doing so would eliminate a major source of surface water contamination, reduce or eliminate the need for evaporation ponds that contain badly polluted produced water that is toxic to wildlife and creates hazardous emissions, and reduce or eliminate the saline water disposal problem. Pipelines deliver crude oil and natural gas around the country. They could also capture produced water and send it back for reuse to most if not all Utah wells. Yet, as of 2014, recycling produced water remains voluntary.

The quick and exponential spread of drilling and fracking, the persistently high oil prices that expedite drilling thousands of new wells, and exemptions from federal environmental standards that safeguard the public suggest to me a parallel between oil and gas operations and other industries that historically impacted Utah's watersheds and water quality. Unregulated or marginally regulated clear-cutting of forests, for instance, led to a series of avalanches that repeatedly destroyed the town of Alta. Uranium mining operations near Moab have left the region to contend with radioactive mine tailings found at the Atlas uranium mine on the banks of the Colorado River near Moab. These contaminated tailings continue to pose a threat to drinking water supplies downstream. All this is by way of saying that the unprecedented boom in oil and gas drilling has created a number of issues that the state and federal government have only begun to address. Utah is beginning to take a number of these issues under serious consideration. On June 26, 2013, regulations governing the management and "disposal of produced water and oil-field wastes" were released.[45] These are a good start if enforcement is carried out, for otherwise the new regulations are useless.

As indicated above, wastewater produced by oil and gas drilling and fracking operations is a form of highly toxic industrial waste. Years ago, when businesses could still discharge carcinogens like polychlorinated biphenals (PCBs) into rivers, few knew about the problems associated with such chemicals. Those who did know either remained silent or were unable to enact timely reforms. Cleanups in Superfund sites across the country have cost taxpayers billions of dollars. If in fact the risk of water contamination is slight, it would not cost much for producers to post large bond sums to address future unforeseen environmental complications that would cover a worst-case scenario. In the case of the PR Springs tar sands project, for instance, the "reclamation surety" is only $1.6 million.[46] This is intended to cover only the costs of reclamation—grading, application of topsoil, and revegetation. What guarantee is there against watershed contamination or damage to the springs and seeps in the area? What protects taxpayers from a firm declaring bankruptcy if the costs of cleanup are too great and future rewards too limited to make a cleanup worthwhile? Establishing regulations to ensure that cleanup costs are covered for companies extracting great wealth from public lands seems to me like a minimum ante.

But political pressure to minimize regulations is significant,[47] as oil and gas operations provide thousands of high-paying jobs in rural regions. Recently I met a 30-something Utahn who works as tool pusher on a drilling rig in North Dakota.[48] He told me that he earns about $170,000 per year. "Not bad for a high school education," he added.

"At that rate," I told him, "you'll be a millionaire in a few years." I was somewhat envious of his salary, which dwarfs my own as an educator.

"Ahh, I don't save anything," he lamented. "My girlfriend had a $7,000 credit card bill last month." And in that regard he's probably not much different than Americans in general, who tend to embrace a live-for-the-moment spirit. It's not surprising that our government embraces similar short-term thinking, which clearly contributes to the pressure to drill. A recent article notes that many Vernal high school students drop out, "lured by the chance to make money working in the oil fields," and then buy "a house, a big truck, some ATVs."[49] I can understand that. In high school I had a minimum wage job ($3.35/hour) at the local garage in my hometown, a full-service Mobil filling station where I pumped gas, checked the oil, fixed tires, and did light repairs when the mechanics were not on

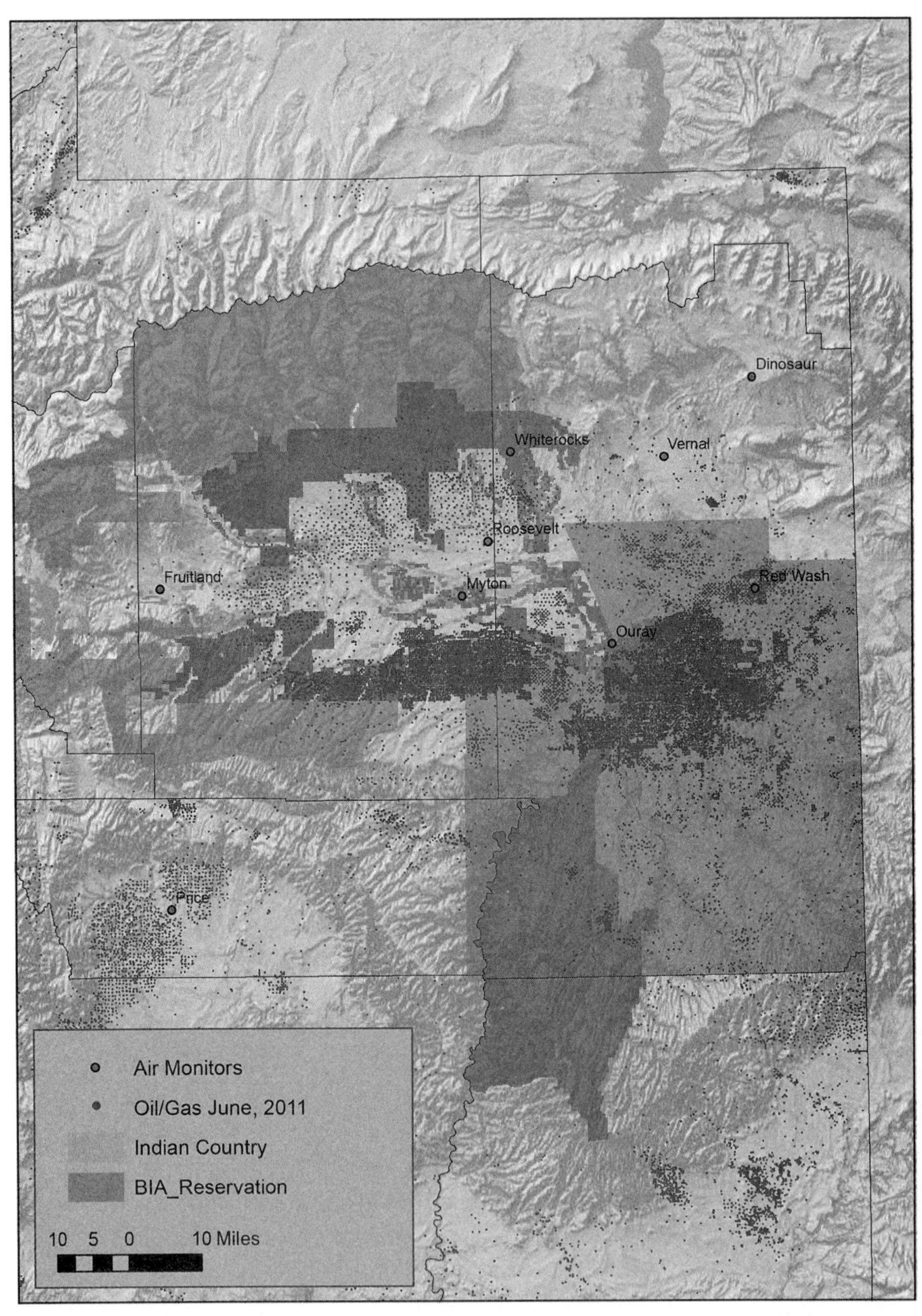

Figure 11.4. Uinta Basin oil and gas wells and areas at risk from air pollution. Utah Division of Air Quality.

duty. Had I worked forty hours per week full-time at this job, my 1983 pay would have come to $6,968 annually. I would have been tempted by work that paid the 1980s equivalent of $170,000 annually. Frankly, that salary is appealing even today. More than once I've considered seeking work in the oilfields or on a construction site in eastern Utah during the university's summer recess.

Such high wages encourage turning a blind eye to problems, denying their existence, or doing a bit of foot-dragging when it comes time to fix them. One major issue connected to hydrocarbon development in the Uinta Basin, for instance, is air pollution. Especially in winters with persistent snow cover, the region frequently does not meet federal air quality standards; in recent years air pollution has exceeded federal limits, often by nearly 200 percent.[50]

The Uinta Basin frequently has some of the coldest temperatures in Utah and suffers from the same thermal inversions that plague the Wasatch Front, some 150 miles west: warmer air aloft traps colder air along the ground, resulting in thick fog. The fog by itself is not toxic. But the emissions from the thousands of oil and gas wells across the basin, the equipment needed to service them and drill new wells, and a coal-fired power plant in the region put pollutants into the air, where they remain trapped. It's a fact that oil and gas wells all produce measurable emissions. According to the Utah Division of Air Quality, "measurements taken around oil and gas operations in the Basin showed that 98–99 percent of the VOC emissions and 57–62 percent of the NO_x emissions came from oil and gas production."[51] With fracking operations, for instance, "at each stage of production and delivery, tons of toxic compounds (VOCs), other hydrocarbons, and methane (fugitive natural gas) can escape and mix with nitrogen oxides (NO_x) from the exhaust of diesel-fueled, mobile, and stationary equipment to produce ground level ozone,"[52] which is considered a serious threat to human health.[53] These emissions, as well as fine particulate emissions known as PM2.5 that consist of "organic chemicals and acids such as nitrates and sulfides," contribute to the basin's worsening air quality problem.[54] It manifests itself in terms of an increased number of asthma cases, emergency room visits due to breathing problems, heart attacks, and more during periods of high pollution.

The good thing about air-quality problems is that bad air is visible, so it gets people's immediate attention. Anyone living in an area with severe air pollution, particularly along the I-15 corridor from Payson, Utah, north through Salt Lake City and on to Brigham City, can see a brownish-yellow haze hanging over northern Utah valleys in the winter and also in summer. Frequently winter air pollution is severe enough to place the Wasatch Front in the top ten listing of most polluted cities nationwide. Sometimes conditions earn the region the dubious distinction of having the nation's worst air quality. The Wasatch Front's 2.1 million residents can see the bad air, smell the bad air, and, unfortunately for all too many, feel the bad air in their respiratory tract as their throat and lungs become irritated by the toxic stew of pollutants. Utah Department of Health data indicate that children living near the refineries in North Salt Lake and Woods Cross, for instance, suffer from asthma rates that are 300 percent of "the normal prevalence." Heart attacks and strokes spike in correlation with pollution episodes that leave many gasping for air.[55] In the last few years a similar set of urban air pollution problems has come to the cities and towns of the Uinta Basin, despite its smaller population (35,000 plus). The cause lies in the thousands of new and existing wells and the equipment needed to establish and service them.

I mention air pollution because the problem threatens to curtail oil and gas exploration and production in the Uinta Basin. If producers are unable to find ways to cut the emissions that are creating the wintertime air-quality issues, it's possible that the EPA will declare the area a nonattainment region and require the state to develop a plan to bring air quality back up to federal standards. This plan could reduce the number of wells drilled each year; in a roundabout way solving the air pollution problem could also have the effect of helping to reduce the pressures on water resources in the region. Industry has been feeling the pressure to clean up the air, though no major action has been taken as of 2014. I think many oil and gas producers understand the water issues—both the use and disposal of water are a concern. With the exception of the toxic-looking evaporation ponds, however, the problem is that the water pollution issues seem invisible to the public at large. This isn't water pollution fifty-some years ago, such as in 1969, when Cleveland's Cuyahoga River infamously caught on fire in an event that helped push the nation toward enacting more stringent laws protecting water quality. It's also not similar to very visible current water pollution problems in Utah that are obvious to the public, such as the vast floating mats of algae—many toxic—caused by nutrient pollution from agricultural operations.[56] Unlike some types of air pollution, water pollution can be invisible, impossible to confirm without sophisticated equipment.

People sometimes say that a person needs to be for something—not just against something. All too often groups advocating for human health or the environment are characterized simply as people who are against something—jobs, say, or the economy in general. But groups advocating for human health and the environmental *are* for something. They are for healthy families and for healthy places to live, work, and recreate. They are for places where people do not have to live in fear of having their air or water contaminated. These are not unreasonable ideas to support.

For now we are reliant on combustible fuels to power the transportation network, heat homes and businesses, and keep agriculture going. Are the environmental costs of current hydrocarbon extraction and wastewater disposal too great? I am not advocating the elimination of oil and gas production in eastern Utah. But for the long-term health of the region and the people who live there, shouldn't the agencies and politicians who regulate producers insist on universal adoption of best-in-practice technologies and procedures in the areas of wastewater management and the composition of drilling and fracking fluids, as a handful of leading companies have? I also question the logic of tar sands and oil shale development that requires huge quantities of water in parts of the state with extremely limited resources. The question of whether fossil fuel production should be ramped up at a time when climate change is having profound impacts on other areas of the state's economy also bears consideration; you can find discussion of climate change elsewhere in this book.[57]

Finally, I question the wisdom of turning the Beehive State into the land of 20,000 wells, with ever-more acreage becoming pocked and scarred with wells and roads. I'm still pondering these issues and others six days later, as we emerge from Gray Canyon, row our rafts through the choppy waves of Swasey's rapid, and pull up to the takeout that will conclude our Green River trip. We de-rig the rafts under the hot sun on a concrete boat ramp, where a Uinta Basin–based commercial rafting outfitter leaves a diesel truck idling on the boat ramp for an hour. Though the livelihoods of the guide and driver depend on the river, their choice to leave the truck's engine running unnecessarily suggests that they are unaware of the pressures that oil and gas development place on Utah's water resources. Will others begin to think about such issues? Can we manage Utah's fossil-fuel industry to ensure clean air, clean water, and a strong economy now and in decades to come? Reading this chapter may help.

Notes

1. Hal Crimmel, *Dinosaur: Four Seasons on the Green and Yampa Rivers.*
2. Leslie Haines, "Uinta Basin," 6.
3. Oil shale, often referred to as kerogen, is a sedimentary rock that must be mined. Oil shale is then either burned directly or heated to about 500 degrees Celsius and refined to produce oil. Shale oil, in contrast, often referred to as "tight oil," uses drilling and fracking technologies to bring crude oil to the wellhead.
4. Haines, "Uinta Basin," 2.
5. "2012 Oil Shale and Tar Sands Programmatic EIS," Department of the Interior, Bureau of Land Management, http://ostseis.anl.gov/guide/index.cfm.
6. Ibid.
7. For more detail on tar sands, see "Fossil Foolishness: Utah's Pursuit of Tar Sands and Oil Shale," published by Western Resource Advocates of Boulder, Colorado, in 2010, http://www.westernresourceadvocates.org/fossilfoolishness/fossilfoolishness.pdf.
8. For a further discussion of the impacts of climate change on Utah, see chapter 5 by Daniel Bedford herein.
9. David W. Pershing and Kerry E. Kelly, "Analysis of CO Emissions from Unconventional Fossil Fuel Resources," Institute for Clean and Secure Energy at the University of Utah, http://www.ices.utah.edu/leftnavid3subleftnavid9subpage9.
10. John Baza, State of Utah, Department of Natural Resources, Utah Division of Oil, Gas and Mining, "Hydraulic Fracturing: Water Resource Impacts of This Oilfield Well Stimulation Technique" (presentation, American Water Resources Conference, Utah Section, Salt Lake City, May 14, 2013).
11. In June 2013 the Utah Division of Oil, Gas, and Mining indicated on its website that Utah has "approximately 4,300 producing oil wells and 6,700 producing natural gas wells." See "Well Counts," State of Utah, Department of Natural Resources, Utah Division of Oil, Gas and Mining, http://oilgas.ogm.utah.gov/Statistics/Well_counts.cfm.
12. "Greater Uinta Basin Oil and Gas Cumulative Impacts Technical Support Document," U.S. Department of the Interior, Bureau of Land Management,

March 2012, 7, 10, http://www.blm.gov/pgdata/etc/medialib/blm/ut/vernal_fo.Par.57849.File.dat/GCW%20Cums%20TSD%2003-22-12%20final.pdf.

13. Ken Kreckel, "Directional Drilling: The Key to the Smart Growth of Oil and Gas Development in the Rocky Mountain Region," 2007, 1–56, http://wilderness.org/sites/default/files/legacy/Directional-Drilling.pdf.
14. Fracking is a process of horizontal drilling into layers of shale that contain oil and gas. Wells are injected under high pressure with a mix of water, chemicals, and sand to crack open tiny pores in the rock, allowing the gas or oil to flow out to the wellhead and be captured.
15. "Introduction to Produced Water," National Energy Technology Laboratory, U.S. Department of Energy, 2013, http://www.netl.doe.gov/research/coal/crosscutting/pwmis/intro.
16. Ibid.
17. Adam Voge, "Plans for 23,000 Wells Buoy Wyoming's Oil and Gas Future," *Casper Star-Tribune*, April 1, 2013, http://trib.com/business/energy/plans-for-wells-buoy-wyoming-s-oil-and-gas-future/article_f3d3ec28-41a5-566c-bde1-3009881dbe48.html.
18. In May 2014, an aging oil well blew out a dozen miles south of the town of Green River, Utah, and spilled thousands of gallons of produced water and oil into Salt Wash, four miles from the Green River. The contaminants flowed three miles before the spill seemed to be contained, just one mile from the river. However, three days later Jim Collar, who was camping downstream from Salt Wash on the rim of Labyrinth Canyon, spotted an oil slick covering the river from bank to bank. Officials later acknowledged that oil and produced water had in fact reached the Green River, a major tributary of Lake Powell. See Kristen Moulton, "Remediation of Utah Oil Spill Continues in Grand County," The Salt Lake Tribune, May 28, 2014, http://www.sltrib.com/sltrib/news/57998688-78/oil-river-green-wash.html.csp.
19. Frank R. Spellman, *Environmental Impacts of Hydraulic Fracturing*, 109.
20. Vikram Rao, *Shale Gas: The Promise and the Peril*, 38.
21. John Rogers, State of Utah, Department of Natural Resources, Utah Division of Oil, Gas and Mining, phone interview by Hal Crimmel, June 28, 2013.
22. Spellman, *Environmental Impacts of Hydraulic Fracturing*, 142.
23. For an example, see http://energyindepth.org/just-the-facts/#groundwater-contamination.
24. A "letter to the EPA from Henry Waxman, D-California, Edward Markey, D-Massachusetts, and Diana DeGette, D-Colorado," accused "14 fracking companies of injecting more than 32 million gallons of diesel fuel into the ground in 19 states between 2005 and 2009." Cited in Abrahm Lustgarten,

"Drilling Industry Says Diesel Use Was Legal," *ProPublica*, February 8, 2011, http://wyofile.com/propublica/diesel-fracking/.

25. See Spellman, *Environmental Impacts of Hydraulic Fracturing*, 141–47, for a partial list of chemicals. The FracFocus Chemical Registry also includes a list of chemicals commonly used in hydraulic fracturing.
26. Lisa B. Uriell, "Analytical Report," Job Number: 280-28076-1, Job Description: EPA Pavillion Fracking (Prepared for U.S. Geological Survey, May 15, 2012), 46–48, http://www.epa.gov/region8/superfund/wy/pavillion/phase5/Appendix9_J28076-1_EPA_Std_Tal_L4_Package_MiniFinalReport.pdf.
27. Mead Gruver and Ben Neary, "EPA Won't Finalize Wyoming Fracking-Pollution Study."
28. *Gasland II*, directed by Josh Fox.
29. Baza, "Hydraulic Fracturing."
30. John Baza, State of Utah, Department of Natural Resources, Utah Division of Oil, Gas and Mining, personal interview by Hal Crimmel, Salt Lake City, May 14, 2013.
31. Edwin Dobb, "The New Oil Landscape."
32. Ibid.
33. In October 2013 a Tesoro pipeline spilled 20,600 barrels of crude oil into a North Dakota wheat field. The company claimed that no groundwater contamination occurred, but the farmer, Steve Jensen, said that the seven acres of land will be unusable for some years.
34. Rao, *Shale Gas*, 32.
35. Baza, "Hydraulic Fracturing."
36. Clive Bates and Andy Rowell, "Tobacco Explained: The Truth about the Tobacco Industry Explained...in Its Own Words," World Health Organization, 7, http://www.who.int/tobacco/media/en/TobaccoExplained.pdf.
37. Quoted in ibid., 9.
38. For instance, a post in "The Real Promised Land" states: "Table salt, swimming pool cleaner, candy, and salad dressing. These are all of the common, household applications of the additives—that make up .5%—in the fluid used in the hydraulic fracturing process." See http://www.realpromisedland.org/.
39. The Groundwater Protection Council's executive director happens to be the former associate executive director of the Interstate Oil and Gas Compact Commission.
40. Jim Polson, "Frac Focus Fails as Fracking Disclosure Tool, Study Finds."
41. Baza interview.
42. Michael D. Vanden Berg, Utah Geological Survey, "Saline Water Disposal Issues in the Uinta Basin, Utah" (presentation, American Water Resources Conference, Utah Section, Salt Lake City, May 14, 2013).

43. Rogers interview.
44. Vanden Berg, "Saline Water Disposal Issues."
45. "R-649-9," State of Utah, Department of Natural Resources, Utah Division of Oil, Gas and Mining, 2013, https://fs.ogm.utah.gov/pub/Oil&Gas/Notices/R649-9_eff_07012013.pdf.
46. "Tar Sands," Tab, State of Utah, Department of Natural Resources, Division of Oil, Gas and Mining, 2013, http://linux1.ogm.utah.gov/WebStuff/wwwroot/division/tabs.html.
47. The federal government's 2005 Energy Policy Act, discussed earlier in this chapter, enacted a host of pro-development policies that politically complicate more restrictive policies at the state level.
48. A tool pusher is the industry term for the foreman of a drilling rig.
49. David Gessner, "How Big Oil Seduced and Dumped This Utah Town."
50. Amy Joi O'Donoghue, "Uintah Basin Ozone Problem Triggers Lawsuit against EPA."
51. "Uintah Basin: Air Quality and Energy Development," State of Utah, Department of Environmental Quality, Division of Air Quality, last updated September 23, 2013, http://www.deq.utah.gov/locations/uintahbasin/faqs.htm.
52. Spellman, *Environmental Impacts of Hydraulic Fracturing*, 210. NO_x are nitrogen oxides, toxic air-borne chemicals typically formed by the combustion of fossil fuels. One of these oxides, nitrogen dioxide, which forms a brownish haze, is "believed to contribute to heart, lung, liver and kidney damage" (ibid., 223).
53. Ground-level ozone is considered extremely hazardous to human health. It's often described as "sunburn on the lungs" and is monitored by the EPA and the Utah Division of Air Quality.
54. "Air Pollutants: Particulate Matter," State of Utah, Department of Environmental Quality, Division of Air Quality, http://www.airquality.utah.gov/Public-Interest/about_pollutants/about_pm.htm.
55. Howie Garber, "Air Quality on the Wasatch Front Is a Public Health Crisis: How to Protect Your Health," Utah Physicians for Healthy Environment, 2011, http://www.uphe.org/general-research/air-quality-on-the-wasatch-front-is-a-public-health-crisis.
56. In 2004, for instance, eighteen cattle died from eating toxic algae. See Amy Joi O'Donoghue, "Utah Faces Polluted Water Woe."
57. See chapter 5 by Daniel Bedford herein.

12

≈

Moving Water

JANA RICHMAN

SEVEN YEARS AGO Patricia Mulroy, the fierce and indefatigable general manager for the Southern Nevada Water Authority (SNWA), announced plans to build three hundred miles of pipeline to gather water from rural valleys in northeastern Nevada and deliver it to the city of Las Vegas, at which point I sat up and took notice. You don't need much water knowledge to know that the aquifers under our feet don't abide by state borders and water rights, so Mulroy's plan includes sucking water out from under Utah. I suppose that's why this particular declaration grabbed my attention. It hit close to home: Utah's west desert.

The summers of my single-digit youth were spent exploring—in both directions and in glorious oblivion—the creek that ran on the far side of my backyard's chain link perimeter. Beyond the creek, open wheat fields housed what upon reflection seemed a rather large population of the horned lizard known to us as horny toads. The front yard represented civilization—lawns, curb and gutter, sidewalks for chalk-drawn games, and paved roads for bicycling. I had my pick of worlds—wild or tame—and indulged in both.

The ditch, as we called it, was about three feet wide and maybe a foot deep in most places. It had a good flow. Into the water my best friend and I dropped all things—plastic balls, sticks, leaves, paper boats, stuffed animals, children's books, and dolls—to check the item's floatability. The water itself held more wonder and entertainment value than the thing we dropped.

Once we detected a good floater, we'd launch it, rush out front to our waiting bikes, and race down the sidewalk several blocks before thrashing through pussy willows at the place where we hoped to intercept floating Barbie before she went under the road and disappeared for good. My enthrallment with moving water has never diminished, though my access to it has.

When I began writing about water, incited by Mulroy's announcement, the immensity of my ignorance on the subject became apparent. It's not much comfort, indeed none at all, to know that I have a great deal of company. I joke about those who believe the kitchen faucet to be the source of their water, but the reality doesn't stray far from the gag, making it more tragic than funny.

I might be less sheepish in admitting the scope of my ignorance had I grown up in the lushness of upstate New York, the swamps of Florida, or the damp forests of the Oregon Cascades, but I grew up in Tooele County—Utah's west desert—although during my youth I never heard it referred to as such. Maybe I slept through geography classes, maybe I was distracted by the closeness of the boy at the next desk, or maybe my senses were dulled by the soothing sound of unfettered water nearby, but I was far into adulthood before Wallace Stegner's slim book *The American West as Living Space* informed me of my desert home and what it means to be a desert dweller.

I don't know whether to fault myself or the education system for the lack of geographic knowledge that followed me through life. Probably both. I stumbled through a K-12 education and four years of college with a bachelor's degree at the end—all of it in the state of Utah—having never heard of Stegner, although he was writing about the American West in general, and my home state in particular, before I was born. Yet he was notably absent from Tooele High School in the early 1970s.

In his introduction to *The Sound of Mountain Water*, Stegner writes, "In most parts of the West (Utah is one exception) a child is likely to learn little in school about the geography and history of the region that is shaping him. He gets them through the pores if he gets them at all."[1] I take exception to Stegner's exception. I am living proof that Utah fit the educational mold of which he writes. What geographical knowledge I've acquired over the years has indeed come through the pores.

One hot summer day between second and third grade, my friend and I hatched a plan to follow our ditch upstream, an adventure of such great proportion that I felt like Nellie Bly. We had no destination in mind; I'm chagrined to admit we had no idea where the water would lead us. Unbeknown to us, we set out in search of the ditch headwaters.

I consider myself lucky to have grown up in an era of mothers who sent their young into the town wilds fully expecting that they would reappear at sundown. I find good reason to wax sentimental about those less fretful times. 'Twas sweet luxury to be beyond a parent's reach for the span of a summer day at the age of eight, pure opulence to discover the earth through my own senses.

We traveled with domestication on our left and wilderness on our right, the creek providing the perfect line of demarcation between the two—a scene that my eight-year-old self took for granted, a static world where creeks and wheat fields don't disappear in two years' time. We hadn't gone more than a mile before stopping for lunch amid willows, having spent most of our energy in the plotting phase of our daring quest. Although the sun was high and hot when we pulled our feet from the creek and laced up our Keds, I suggested that we turn back. Better safe than sorry—one of the worst slogans ever set in type—was my early philosophy of life. But my friend, the alpha female of our two-person team, wouldn't hear of it. We pressed on. I was considerably out of my comfort zone when we reached the cement plant on the south side of town and near the point of panic upon reaching Highway 36 just beyond. We crossed the highway and paused at the wrought-iron fence erected around a few pioneer graves, including those of several children, in the mouth of Settlement Canyon, an arrival so portentous in my mind that I found myself near tears. Immune to my histrionics, my friend kept going, and I stumbled after her.

As we walked directly into the Oquirrh Mountains, a strange and lovely thing happened: the ditch, *my* ditch—the one that carried toys away and disappeared under paved roads and into concrete culverts—turned itself as if by magic into a mountain stream, a beautiful, gurgling flow that led us deeply into shadows of large cottonwoods, a cool darkness so enchanting that it stripped me of my breath and fears.

Less than a mile into the canyon we stopped at a wide place in the creek, a canopy of branches and leaves above our heads and a rough plank bridge across the water built seemingly just for us. We dangled our feet, waded and splashed, padded barefoot on the dusty banks, and lay on our bellies to slurp our fill of clear mountain water before starting home as the sun dropped.

It was my first serious romance with a natural place, a place that provoked yearning and dreaming, a place that churned the belly and dizzied the brain and unveiled a world of possibility. It was the first time I understood that open air, dirt, trees, and flowing water make up the natural habitat of humans.

We returned many times that summer and the summer to follow. We planned our lives, chose our husbands, named our children, and whispered our confidences as if the place itself would grant each fantasy and protect every secret. We had slipped through a secret chamber into our own enraptured forest.

Then it was gone. Bulldozed, stripped, destroyed, and flooded. In 1966, the year I turned ten, I witnessed my first western water project: an earthen dam constructed in the mouth of Settlement Canyon. I had heard my father—a man already in the plotting stages of buying a small ranch, a man who loved the sight of water spurting from metal pipes as much as I loved it cutting through the earth, a man who in future summers would give new meaning to the phrase "moving water," as in "we need to get that goddamned water moved"—talking excitedly about the dam. But never in my mind did I align the construction of the dam with the destruction of my place until it happened.

Certainly the flooding of the mouth of Settlement Canyon is a tiny blip of a thing compared to, say, the flooding of Glen Canyon, which happened around the same time. But my ten-year-old heart shared the depths of despair described by David Brower as he watched Glen Canyon fill with water in the years after the Glen Canyon Dam gates closed in 1963: wretched desolation for the loss; unfathomable disbelief that so many could accept it so easily.

As a sixth-generation Utah Mormon, I come from a long line of master irrigators, a great heritage of water movers. When Brigham Young said, according to the myth, "this is the place" and shortly thereafter promised to "make the desert bloom like a rose," in true Mormon fashion, he girded his loins and went to work. Under his command Mormons moved earth, built ditches, and took water from mountains to valleys, insisting the desert produce crops that it had never before considered while creating what would turn out to be one of the largest and most successful irrigation systems of its time and possibly the inspiration for many water projects that followed.

Every "water project," every large or small manipulation of water from its natural course, carries with it a loss. Whether it happens through damming and flooding, through pumping and draining, or through the urban development supported by such projects, a natural place is forever changed. I knew nothing of the reasons behind the Settlement Canyon Dam at the time it was built. I only knew it had stolen *my place*, and to this day the tiny, still reservoir, accessed by a recently added tollbooth and gate, strikes me as one of the most repulsive places on earth and becomes more revolting and unsightly with each visit, as if the dam set into motion a decline from grass, trees, and water to concrete, beer cans, and used condoms. Promotional materials for the area publicize the reservoir's "unsurpassed beauty" and recreational fishing opportunities, but beauty is a relative thing. You can see the beauty in a human-made body of water nestled into a valley only if you never felt the spirit of the place snuffed out by it. I suppose that Lake Powell is beautiful to some who never saw what lies beneath. I am one of those who never visited Glen Canyon before the flood, but I've spent the last four years wandering the desert near Lake Powell, and my ten-year-old heart identifies the loss.

In addition to Glen Canyon Dam, I did not notice other major water projects in my province and lifetime, including the massive and controversial Central Utah Project. The work of the Central Utah Project spanned more than seven decades, during many of which I was old enough to pay attention. But I didn't. Still, I've probably seen most every dam within a three-day driving distance of Tooele. My father, giddy with dams, often coaxed us into the car to "go see the dam," whether that meant a short drive to check the water level in Settlement Canyon or an 1,800-mile drive to see Glen Canyon Dam, Hoover Dam, Davis Dam, and Parker Dam during one car trip. I was duly impressed with the constructing prowess of my species. I don't know how common it was in the 1960s to create vacations around visiting dams, but for us Glen Canyon Dam was the destination, the Grand Canyon a side trip.

The Settlement Canyon Reservoir, which collected and stored snowmelt and a spring-fed creek, not only took my sacred place but also ended my joyful play in open ditches, which soon thereafter disappeared, followed shortly by open fields. Once we had a modern water system, we could grow the town, irrigate farmland, and water lawns without restraint. I've never met anyone else who felt that as a loss rather than something to celebrate. *Without restraint*, however, was a short-term dream now seeking new dreamers, a commonality among western water projects.

Once I left Tooele for good in the early 1980s, I spent many years running from Utah's west desert, a place that had trapped the awkwardness and pain of my adolescence in its arid severity. The west desert held the yearning part of me, the part that could never be mollified because its object eluded me. One evening in a reading group discussion of Julie Otsuka's novel *When the Emperor Was Divine*, set in the Topaz Internment Camp in the west desert, the conversation turned to the dreadfulness of the place—not the camp but the desert itself. All generally agreed that a more horrifying geographic location could not be found on earth. I made a comment that the west desert has been devastated by off-road recreationalists and the military, and the response from one and all was, "Who cares?" Evidently I do. My defenses escalated and tears brimmed my eyes, although I had been asking myself the same thing for years. Who cares? I hadn't driven in quite some time any farther than the west end of Vine Street in Tooele, where my parents still lived, busy as I was transforming myself into a city girl, but apparently the west desert still had hold of me.

The next day I drove past Great Salt Lake, through Tooele, and past Settlement Canyon Dam. I continued through Stockton and Rush Valley and over Johnson's Pass on Highway 199, threading my way between the Onaqui and Stansbury Mountains. When the guard gate of Dugway Proving Grounds appeared, I took a left off the pavement onto the Pony Express Trail and stopped at Simpson Springs, a place familiar to me from a few high school keg parties. There I spent the day alternating between joy derived from sheer beauty and solitude and sadness resulting from exactly the same.

To enter the west desert is to rake the rawness of the most sensitive spot I carry. Everyone has one, a tender place inside—the place of the child, the place of loneliness and vulnerability. That's the west desert. It never gets old; it never gets easy; it never stops breaking my heart. Driving into it, walking through it, sitting on the side of a gravel road to watch a band of wild horses graze gives me a simultaneous rush of peace and sorrow; tears of solace and tears of regret flow in unison.

The west desert has taken a lot of abuse from humans: nerve gas spills, nuclear waste storage, and off-road vehicle destruction, to name a few. And although I have empathy for ranchers, having grown up as the daughter of one, I acknowledge the damage done by the establishment of agriculture in a place that gets less than seven inches of rain a year, including the disappearance of springs, native plants, and wildlife in favor of hay and cattle production. But there remain places—Fish Springs comes to mind—that preserve the desert's still spirit and vibrant ecosystems. And the residents and ranchers of some towns—Ibapah and Callao— live in relative balance with their desert geography.

With the announcement of the pipeline, I attempted to learn the intricacies of water, but my ignorance spreads like ditch water turned into a flat field. The issue of western water pulls in the complexities of nature, science, politics, law, history, economics, and psychology, and swirls them together in a deep bog. You soon become mired in the muck.

Environmentalists have their lawyers and lobbyists and scientists, cities and states and counties and water authorities have theirs, and agriculture and mining industries have theirs. The scientists' points of view often contradict one another. The politicians choose whatever piece of science fits their agenda. They opt to use one piece of science for one part of the agenda and a contradicting piece for another part, trusting that the convolutions of the issue will keep all interested parties conveniently confused and ignorant, a tactic that works quite well.

The plan for a pipeline project of the Southern Nevada Water Authority (SNWA) has a table of contents alone that runs four pages. If that's not enough to elicit groans, another page adds a list of more than thirty acronyms to learn before wading into the morass, and another page lists thirteen federal and state government agencies with an interest in the project, though not a single one is from the state of Utah.[2]

The history of water in the West is driven by economics and fraught with backdoor deals, greed, power struggles, and gargantuan foreseen and unforeseen—and mostly permanent—consequences. When it behooves us to trot out history we do so, which is what has given the agricultural industry a stronghold on western water for many years. And when it behooves us to ignore history—depleted aquifers, raised salinity content in freshwater rivers, disappearing native fish, invasive species, and the drying up of springs and native plants—we're quick to dismiss it. History also drives psychology, as is the case in the small town where I now live. In Escalante, Utah, we run out of irrigation water every summer regardless of winter snowpack and despite the enormous cost of a newly constructed dam. Water users here operate under the "use it or lose it" policy. We water as if we are one of ten stepping up to a dinner table set for four. We take a fighting stance and get as much as we can as fast as we can before it is gone, which, of course, guarantees that it will be gone soon, which in turn affirms the approach. Those who operate under this popular maxim think that water conservation is a great idea—the more the neighbors conserve, the more I get.

Mulroy claims that she can pump water out of western Utah and eastern Nevada basins without affecting wetlands, springs, lakes, or ranchers' livelihoods. Few believe her. None with the most to lose believe her. Many scientists, including some initially hired by SNWA—former employees of the Nevada office of the U.S. Geological Survey—to produce and run models to predict the consequences of the pipeline, don't believe her. I don't believe her. No one who looks at the history of groundwater pumping projects—Owens Valley in California and the Ogallala Aquifer in the Midwest provide two good examples—would find reason to believe her. I'm not even convinced that Mulroy believes herself. But as in the case of so many water project proponents by the time the consequences catch up to the action, she'll be long gone and she knows it.

Mulroy bristles whenever someone compares the SNWA pipeline project with other groundwater pumping projects such as the early twentieth century Los Angeles water grab from Owens Valley or the midwestern states pumping from the Ogallala Aquifer, but they are worth considering as comparisons. At one time midwestern ranchers saw the behemoth Ogallala Aquifer that sprawls beneath eight states and initially contained an estimated million–billion gallons of water as the answer to their dust bowl–ending dreams. And for a while it was—they turned the plains green. Acting under the "use it or lose it" adage embraced by many, the states and ranchers dropped hundreds of thousands of wells and pumped essentially without restraint. While we like to think—because it makes us happy to do so—that water is a renewable resource, it is not always the case. The water in the Ogallala Aquifer is nonrenewable water—ancient water formed in prehistoric times over thousands if not hundreds of thousands of years—that resides mostly under a water-impermeable surface. In other words, the midwestern ranchers are not irrigating with rechargeable groundwater so much as they are mining a vein of groundwater. Once it's gone, it's gone.

In the 1990s midwestern ranchers began to notice a decline in the water coming out of their wells. Some wells, irrigation systems, and fields had to be abandoned altogether. According to a July 2012 article in *Harper's Magazine,* a study funded by the U.S. Department of Agriculture and headed by Kevin Mulligan at Texas Tech University found that water in the Ogallala Aquifer will be essentially gone in about fifteen years from now, and in many places it is already too low to pump cost effectively.[3]

Mulroy would be quick to point out that the comparison with the Ogallala Aquifer is inaccurate because mining water is illegal in Nevada and Utah: SNWA is asking only for unclaimed rechargeable water. That's where the science and politics get messy. To put it in a simplified way, the water in the valleys identified for pumping by SNWA is a mix of nonrenewable fossil water contained in the Great Basin Aquifer and groundwater that gets recharged from mountain snowmelt. The fossil water pushes up against the renewable water, creating a series of springs, spring-fed lakes, and a shallow groundwater table essential for both agriculture and wetlands. Bald eagles soar above valleys, snowy egrets skitter in for watery landings, and white-faced ibises wade in grassy marshes. Ranchers and environmentalists claim that this delicate balance operates on the slimmest of margins. Some hydrologists have argued that overpumping the groundwater would cause the native flora—particularly greasewood—to dig its roots deeper, essentially "mining" the Great Basin Aquifer.

But the more popular comparison to the Las Vegas pipeline—one that Mulroy is asked about often—is Owens Valley. Consequences from pumping water from Owens Valley to Los Angeles through the aqueduct completed in 1913 included the drying up of Owens Lake and more than fifty miles of Owens River, the disappearance of native flora and fauna throughout Owens Valley, farm and orchard crop failures, and massive traveling dust storms that contributed not only to further ecological damage but to severe respiratory health problems for humans and animals. Mulroy claims that the same sort of thing won't happen with this project because, she says, current environmental laws won't allow it. But many of the hydrologists and geologists involved in the Las Vegas pipeline project from the start—some hired by SNWA and some hired by ranchers and environmentalists—say it is not only possible that the Great Basin could suffer consequences similar to those of Owens Valley but probable. To argue against the idea that greasewood roots shooting deeply in search of water would amount to illegal water mining, SNWA proposed pumping heavily in the initial stages to kill the greasewood—one of the major plants in the basins that keeps dirt on the ground and out of the air.

Scientists don't all agree on "groundwater models" in the basin and range, on how water does or does not flow between valleys, or on where and how quickly it recharges, all of which affect the predictability of the consequences of pumping. In other words, what we know may be different than what we contend we know. Most scientists do agree, however, that monitoring what happens aboveground will not matter once pumping begins. By the time we are able to measure negative impact on the surface in springs, lakes, and wells, the damage will already be done below the surface. Yet this is precisely the condition upon which SNWA was granted pumping approval by the Nevada state engineer—that it would monitor springs for negative impact. According to a study done by John Bredehoeft, formerly the regional hydrologist with the U.S. Geological Survey responsible for water activities in eight western states, it could take centuries after shutting down the pumps before the negative consequences above ground slow and stop, depending on many factors such as where a well or spring is located and the size of the aquifer it draws from.[4] And it could take millennia to reverse the damage done. Bredehoeft's model showed only 60 percent flow recovery 450 years after pumping stopped,[5] and studies compiled by Great Basin Water Network hydrologist Tom Meyers predicted a recovery period of 2,000 years in Cave Creek, Dry Lake, and Delamar Valleys.[6] A 2012 study done by four professors in Brigham Young University's Department of Geological Sciences confirmed that recovery could take millennia or longer and exacerbate threats to spring-fed wetlands already stressed from climate change.[7]

That accounts only for the damage done by pumping alone. According to SNWA's project development plan, in addition to three hundred miles of pipeline, the proposed project includes construction of five pumping stations, three pressure-reducing stations, six regulating tanks (each with a twenty-foot radio antenna on site), a power line sporting steel power poles every six hundred to eight hundred feet along the pipeline route, two primary and five secondary electrical substations, and more than 225 miles of new roads plus heightened use of existing roads. In addition to these permanent structures, construction will require eight borrow pits (each approximately seven acres in size), ninety-seven three-acre construction staging areas—one about every three miles along the pipeline route—construction support and storage sites, and construction camps for workers. The plan also allows for permanent "future facilities," which remain somewhat nebulous but could include from 144 to 174 wells along with the "electrical facilities, heating, ventilation, air conditioning equipment, and control facilities" required to operate them and access roads required to maintain them, from 177 to 334 miles of "future collector pipelines," additional pumping stations, additional power facilities, power lines, and access roads, the number of which cannot yet be accurately determined. According to the Great Basin Water Network, the current estimated price tag facing Las Vegas residents is about $15 billion—the most expensive water that anyone has ever watched swirl down the drain.[8]

The plan vows to do its best to disrupt, maim, and kill as few "sensitive species" as possible. It claims to provide adequate space between power conductors in hopes that a golden eagle can fly between conductors without injury and will try to "discourage birds from flying between conductors" by using a "230kV pole conductor support design."[9]

Whether or not the pipeline ends ranching and dries up wetlands and springs remains uncertain only in a few politically driven minds. But the devastation of the basin and range as we know it—from Utah's Snake Valley in Tooele, Juab, and Millard Counties to Nevada's Spring Valley, Cave Valley, Dry Lake Valley, and Delamar Valley—is not in doubt in most rational minds. The landscape, the ecosystem, the native plants, the open space, the roadless desert, the wildlife, the insect patterns, the peace, the quiet, the undisturbed solitude—forever gone from the installment process alone before a single pump gets turned on. The question is not whether the specified area of basin and range will be changed, but whether anyone cares—or whether enough care.

A lot of snarling has taken place in the years—and continues still—since the pipeline announcement, as will happen when animals compete for scarce life-sustaining resources. One rancher from Callao, Utah, declared the pipeline a moral issue by proclaiming in at least one interview that the water shouldn't go to support lives of "glitter, gluttony, gambling, and girls," as contrasted with the apparently more deserving rural folks "personified by cattle, children, church, and country." Mulroy has growled in front of a camera that "they can't spell conservation in Salt Lake City."[10] Both statements hold a sliver of truth deeply buried in a slice of sanctimony.

According to figures from the Great Basin Water Network, Utah's water conservation (as indicated elsewhere in this collection) is indeed abominable compared to that of Las Vegas and other western cities.[11] In Utah we've internalized the directive to "make the desert bloom like a rose" as if it were God's eleventh commandment.

I agree that construction of a 300-mile pipeline to pump water from its natural course is indeed a moral issue, although not, as the Callao rancher would have it, one based on the goodness of residents in either place. I don't believe that a rancher living in Spring Valley necessarily holds the moral high ground over a card dealer living in Las Vegas. I assume that the residents of Las Vegas are as devoted to their children, church, and country as are the residents of Callao. I see it as a moral issue based not on the character of rural versus urban residents so much as on the character of the human species in general, a moral failing in refusing to acknowledge that we are not—and never will be—nature's controllers, that we are not a superior species here to be served by all others.

Water has its own nature. It cares not for the needs and whims of humans. It is not dependent upon us; we are dependent upon it—for our very lives—as are most other species. You might think that would make us humble in the face of nature and careful to maintain the balance that our lives depend upon. "How does man intervene?" Mulroy asks and answers her own question this way: "The magic lies in managing that basin."[12]

We are a strange, arrogant animal believing in our own cleverness in the face of evidence to the contrary. From the moment we began manipulating water in the name of civilization, we have been rushing headlong toward the inevitable: a time and place where human cleverness runs out—although it seems human hubris knows no such boundary—and nature pulls us up short. We have arrived.

"Living in harmony with nature" has become nothing more than a marketing line for "green" products. We have not lived in harmony with nature since the onset of civilization. Instead we believe "the magic lies in managing" nature. We yank water from its natural course through pipelines, dam it, hold it behind dikes, divert it, waste it, pollute it, exhaust it, then move on to our next clever act. A fool's game. In the end, nature wins. Nature always wins. Floods. Droughts. Dust storms. Hurricanes. Tornadoes. Wildfires.

We talk about what we're leaving for our kids and our grandkids, but our actions show that we don't care at all. Not one bit. We talk endlessly about planning for the future—in fact we now *preplan*—while we simultaneously talk about living in the moment, but we're confused about both of those things. Our idea of future planning is to accumulate and hoard rather than conserve and change, and our idea of living in the moment is to blow through what we've accumulated and hoarded.

Our short-term goal—the only kind we have—is to save not only ourselves but also our lives of abundance and convenience. No matter where we fall on the egregiously wide continuum of "haves" and "have nots" in the United States, we are loath to give up what we have—be it a little or a lot—possibly because we have created a world in which the "haves" end up accumulating whatever the "have nots" sacrifice. Little or none of it goes into the future generation fund.

Water projects have allowed western cities to grow beyond their natural capacities and allowed agriculture to flourish in places where agriculture should not exist or should exist only on a small scale. Las Vegas, Salt Lake City, Phoenix, Tucson, and many other cities have grown unrestrained in the middle of deserts that demand restraint. Now what? It is difficult to shrink a city whose economy is based upon growth—as are all U.S. economies. It is difficult to close down a city and disperse its residents. It is difficult to prevent people from moving to the West, although the West cannot support more people. It is difficult to ask people to stop having children because we have no resources—especially water—for those children. It is difficult to demand that ranchers stop ranching, stop doing what they love, stop living the only way they know how to live, the only way their families have known for generations, although it makes little sense to grow hay and graze cattle in the driest states in the nation.

Water projects allow us to live beyond our means, something most Americans enthusiastically embrace. "Live Large!" say advertisers, lenders, corporations, universities, governments, parents, and churches. We take the idiom literally. The idea of living where we are, in the small space we occupy, is as foreign and horrifying to us as hauling a bucket of water from a creek for daily use. At the very least we hope to send our children to college. No one could argue that as a worthy goal. For what is life without a formal education and how else will we get a good job? We must have a car and a second car or we can't get to our jobs—which are nowhere near our homes—and we need those jobs to afford the cars and the education and a house with multiple bathrooms and possibly a second house to get away from the first house and the demanding job. Because we don't like our jobs but must spend an enormous amount of time doing them so we can live large, we need distractions for our sanity and convenience: TVs, computers, cell phones, and transatlantic flights taken in the name of enrichment and cultural education. We need access to coffees grown in Colombia and tea grown in China and fresh produce in the middle of winter, and those of us who have broadened our horizons through education and travel need saffron grown in Spain. We must have at least two children. And they must have a car and a house and an education and children and coffee and fresh produce in the middle of winter. And this lifestyle won't run out because it can't run out because we *need* it not to.

Ranchers like to remove themselves from the "living large" category because they stay put, work in the weather, and are often land rich and cash poor. But they too live a life of privilege, be it a pastoral one instead of an urban one. The luckiest live on inherited land purchased when land was cheap, using up to 80 percent of the water in some states to maintain their lifestyles and reaping the benefits of public land grazing permit rates based more on cowboy tradition than on current market value, while simultaneously celebrating soaring market prices of beef and hay. Given the opportunity, how many of those making $18 an hour working in the city of glitter and gluttony wouldn't trade places with the rancher, including those ranchers who work a nine-to-five job to support their ranching lifestyles?

As I see it, here's the truth about the Las Vegas pipeline: even with extended conservation efforts and a recession-caused drop in growth, Las Vegas does need the water or will in the near future. Residents of the valleys to be drained also need the water, as do plants, waterfowl, birds of prey, and the rest of native wildlife in the basin and range. Las Vegas is not the first western city to face a water shortage and won't be the last. The United States is not the first country to face fresh water shortages and won't be the last. There are too many people in the West and too little water. There are too many people on earth and too few natural resources.

When I talk to people involved in western water issues, I ask about population control. The question is invariably met with silence. It is the fat, lumbering elephant in the exceedingly narrow room, a political pariah, a conversation ender. Environmentalists won't discuss it for fear of being thought of as "antifamily" or "anti-Christian." Politicians won't bring it up for the same reason. Yet addressing, or at least discussing, the outbreak of humans and the ability of the earth to support the species is not only the obvious answer but possibly the only answer. After all the complex elements of science and politics and law are done processing and finessing the details, one simple truth remains: if we use water faster than it can be replenished, we die.

The earth has finite space and finite resources upon which life depends. Yet that truth seems difficult for humans to accept. Our inability to hold that truth ultimately leads to snarling fights about water pipelines. Maybe our psyches won't allow us to hold that truth. We are animals with instincts to procreate. Although we have set aside many of our natural instincts—the ones that allow us to see ourselves as part of nature instead of nature's managers, for instance—we apparently cannot set aside the instinct, or religious mandate, to reproduce. We may not be capable of voluntarily shrinking our population to live within our natural means. Our instinct for individual creation obviously overrides our instinct—assuming we have one—for preservation of the whole, so maybe we simply continue down that road until we can go no farther. Our grandchildren might be okay; they might have some road still to travel. Their grandchildren may run out of road, but we don't even know them. Our great-great-grandchildren are hypothetical—we can't *feel* them—which makes our compassion for them equally hypothetical.

E. O. Wilson once hypothesized that humans are predisposed to seek habitats similar to "those in which our species evolved in Africa during millions of years of prehistory,"[13] which includes prominences overlooking a parkland and water, consistent with the savanna hypothesis of human evolution. Those with financial means tend to build their homes on hills or in penthouses, near greenery—golf courses, parks, or meadows—with a lake or ocean view, which would support his hypothesis. For those of lesser means planting lawns and water features will have to do. Here in my small town land around homes was cleared of its native desert flora decades ago and replaced with grass and irrigation systems. Clearing the land was no doubt easier than restoring it. I have neither the capability nor the financial means to return my little plot to its natural habitat, so I'm stuck with either watering or inviting invasive pigweed to take over. I've settled on a combination of the two. Left to its own accord, the desert may eventually restore itself, but it wouldn't happen in my lifetime.

I readily admit, even as a desert dweller (especially as a desert dweller), that I find greenery soothing. I live in and hike the canyons of Escalante Grand Staircase National Monument. When I drop off slickrock into a deep ravine such as Death Hollow to find mossy-walled springs, tall grasses, and tangled vines of poison ivy lining the creek, the tranquillity that enters my body cannot be shrugged off as a luxury. I *need* water not only to quench my thirst but to feed the anima of my being. That cannot happen without encountering water in its natural state, something that we are quickly losing. If I follow the creek down Death Hollow, it will take me to the natural confluence with the Escalante River. If I follow the Escalante River, it will *not* take me to the Colorado River, as it naturally should. Instead it will take me to the backed-up bathtub of Lake Powell. How far could someone travel on any western creek or river before butting up against a water project?

My body provides the mechanics of my life, but it does not operate independently of spirit. They are inseparable. I'm not emotionally or psychologically prepared for a world where all water is captured solely for survival of the body. Each summer I make several trips to Upper Calf Creek Falls to swim in natural water, and each summer the number of people doing the same—possibly desperate for one chance at a disappearing opportunity—increases. Their faces reflect sheer joy and wonder, emotions common when the physical body is immersed in its natural habitat. What happens when this is no longer possible? What if, in our cleverness, we find a way to survive physically without places like Death Hollow and Fish Springs? Say we tap all sources of water the moment—or better still, before—they emerge from the earth and collect and store them for physical human survival as our western populations continue to grow? What happens to a species that carries a genetic imprint of nature when we pervert nature, when children can no longer splash in a creek, when the only place to see moving water is in the dancing fountain of the Las Vegas Bellagio?

I don't know the answer to that question, but it seems that we are determined to find out. My gut feeling is that we lose an essential part of ourselves, the balancing part of humanity, the part that allows us peace in spite of fear, and joy amid sadness—the only part that allows us to love one another.

I'm told, as I write about the water issue, that people are tired of doom and gloom—that if I insist on taking a stand against the Las Vegas pipeline I should propose "feasible alternatives," one of our favorite goal-driven phrases. We're looking for rational solutions, I'm told. Humans cannot conceive of a place or a moment without answers, yet that's precisely where we are. We can stomp our little feet and wave our little fists and demand otherwise, but nature is indifferent.

I don't have answers for water shortages that we will surely continue to face in the West during my lifetime and beyond. I have no faith in the cleverness of humanity and no reason to believe my species should escape the path of any species that destroys the ecological balance of its habitat. Some remain hopeful. Life is not worth living without hope, they tell me, and so they simply choose it. No matter what passes through their field of vision, no matter what information the earth presents to them, they remain insistently hopeful. In my mind that's the definition of delusion, but it may also be an effective coping mechanism.

At the opposite end are those who feel utterly hopeless in the face of the consequences caused by our own actions. They find themselves buried by existential sadness.

Are those our only two choices? I see a lot of space between the two opposites—space for meaning, space for community, space for love, and space for peace. At one time those were empty concepts to me, words flung about by the confused and desperate—and I was, and might still be, one of them. Throughout my life I've run from the devastation and garbage that we've heaped around us—nerve gas in Tooele, sprawling consumption of the Sonoran desert in Tucson, and most recently deadly air in Salt Lake City. Running is not a feasible solution for the masses, and many would call it the coward's way. So be it. I am the person who drops back from the frontline battle with her head tucked low.

I love my home state of Utah, arguably the most beautiful place on earth, but our destruction of it fills me with despair. Still, when I returned after twenty years away, I vowed never to live elsewhere again, and I won't. But while living in Salt Lake City my husband and I found ourselves shrouded in a profound sadness as heavy as the filthy air in our throats and nostrils. Many of the wild places along the Wasatch Front and Back that we had, in our youthful ignorance, taken for granted are gone by way of growth and progress, just like my place in Settlement Canyon.

We fled to Escalante to escape our losses, and here I found those previously empty concepts—meaning, love, peace, and, above all, community—fully defined in a way that I've never before known existed and certainly did not know I needed. I don't know the answers to the problems we've created. Humans can and have survived without electricity, gasoline, and air travel, but we have never survived without fresh water, clean air, topsoil, and nontoxic food—the very things we've sacrificed for the others.

I wish we could find it in our collective hearts to live more benignly among our own and other species, if for no other reason than self-preservation, but I see little evidence of it. I cannot put my faith in politicians, strategists, government officials, and so-called leaders—the Patricia Mulroys of the world—so I put my faith in the individual who believes in love and peace and living in place among neighbors. Some may scoff at that—I once did—but I believe that's where our answers lie.

I put my faith in the individuals who are unafraid to let loose the grasp on material objects in favor of community. I can't say that I'm one of them yet—I'm hardwired for holding tightly to what I have—but I have met them and placed myself among them. And they alone provide more hope than any $15-billion, 300-mile pipeline and its feasible alternatives could ever supply.

Notes

1. Wallace Stegner, *The Sound of Mountain Water*, 9.
2. Southern Nevada Water Authority, "Southern Nevada Water Authority Clark, Lincoln, and White Pine Counties Groundwater Development Project Conceptual Plan of Development, March 2011," http://www.snwa.com/assets/pdf/wr_gdp_concept_plan_2011.pdf.
3. Wil S. Hylton, "Broken Heartland."
4. John Bredehoeft, "Groundwater Monitoring for Mitigation: Will It Work?" Great Basin Water Network, January 2010, http://greatbasinwaternetwork.org/pubs/GROUNDWATER-MONITORINGFOR-MITIGATION_-WILL-IT-WORK.pdf.

5. Ibid.
6. Steve Erickson, Great Basin Water Network, telephone conversation, 2013.
7. J. Gillespie, S. T. Nelson, A. L. Mayo, and D. G. Tingey, "Why Conceptual Groundwater Flow Models Matter: A Trans-Boundary Example from the Arid Great Basin, Western USA."
8. Southern Nevada Water Authority, "Southern Nevada Water Authority Clark, Lincoln, and White Pine Counties Groundwater Development Project."
9. Ibid.
10. Patricia Mulroy, interview, KVBC Television (April 2010), http://www.ksl.com/?nid=148&sid=10387603&pid=1.
11. See figure 6.1 in chapter 6 by Zachary Frankel herein for a graph of western cities' water consumption.
12. Patricia Mulroy, "Extended Interview, Desert Wars: Water and the West," KUED (September 2006), http://www.kued.org/productions/desertwars/mulroy_patricia.php.
13. Edward O. Wilson, *The Creation: An Appeal to Save Life on Earth*, 66.

13

A New Water Ethic

DANIEL MCCOOL

The Nature of Ethics and the Ethics of Nature

ETHICS, as a philosopher friend put it, is "really messy." It is subjective but also influenced by empirical observation and a sense of common vision. It is sometimes fickle and subject to change, evolving over time. Although some ethical considerations are viewed as timeless ideals or eternal truths, some behaviors once considered ethical are now widely assumed to be overtly immoral.[1] For example, it was once considered entirely ethical to deprive women and people of color of the right to vote. It was once considered ethical—indeed recognized in the U.S. Constitution—for white people to own black people, who were not seen as actual human beings. But with time, persuasion, and occasionally violence, the consensus regarding what is right and wrong changes and society develops a new set of rules. As the biologist Edward O. Wilson observed, "Ethics is conduct favored consistently enough throughout a society to be expressed as a code of principles."[2] To Aldo Leopold, ethics "are possibly a kind of community instinct in the making."[3]

In addition to changing materially, the application of ethics in Western society has expanded far beyond the original notions of the human condition and now encompasses other species—indeed the whole planet. This embrace of nature is commonplace in indigenous cultures but a fairly recent development for modern Western societies. Humanist Albert Schweitzer described it well: "Ethics is in its unqualified form extended responsibility with regard to everything that has life.... To be ethical is to share aspirations of the will-to-live."[4] Sandra Postel applies this concept directly to water: "Grasping the connection between our own destiny and that of the water world around us is integral to the challenge of meeting human needs while protecting the ecological functions that all life depends on."[5] This evolution from an anthropocentric to a more biocentric concept of ethics has driven much of the recent discussion regarding water and ethics.

The challenge of constructing an ethical regime regarding water is peculiarly complicated. It is absolutely essential to all life—both human and nonhuman. As a result, water has always provoked emotion and controversy; the old adage that whiskey is for drinking and water is for fighting complicates, if not the construction of an ethical framework, then certainly its implementation.[6]

In addition, water is a quirky and ethereal natural resource that fails to adhere to the dependable, regularized condition that fits so well into a human-engineered legal and economic framework. It falls from the sky or bubbles up from the ground irregularly, and then, with great annoyance, moves from one political jurisdiction to another. It is never created or destroyed, but it seems there is nearly always too much of it, or not enough of it. The elusive goal of "water taken in a little moderation," as Mark Twain put it, is remarkably difficult to achieve.[7] Water in its purest form is the elixir of life, but when tainted it kills off its host. If it goes down your gullet it will hydrate you; if it goes down your lungs it will drown you. Water is in a continual struggle with those who want to dominate nature and engage in what I call "water hubris," that arrogant assumption that humans can control water, which is disproven every time there is a flood or a drought.[8]

The idea of an articulated water ethic is fairly recent. In our traditional thinking, water had a strictly functional role, like a box of nails to build a house or a slicker to keep the rain off. But in the last two or three decades we have come to appreciate the full complexity of water for our culture, our values, our economy, and the planet that keeps us alive. Water is everywhere, if not physically then metaphysically. Its presence in thought, law, and culture seems to vary inversely with its plentitude; Bedouin culture has more words for rain than, say, your typical midwesterner with forty inches of annual precipitation. And even though the academic and philosophical debate over a "new water ethic" is a contemporary phenomenon, every religion on the planet has long employed the metaphor of water. Many "new" ideas about water management have their origins in ancient law and practice.[9] (In short, a water ethic is multidimensional, old yet new, and reflects the passions of the day.) David Feldman writes that "'ethics' as employed in discussions of water resources...has a meaning beyond that of axiology and morals—spiritual and even cultural notions of 'right' and 'wrong' have often been invoked in debates regarding water ethics."[10]

The discussion grows even more vehement when we focus on water use in the Intermountain West, a region characterized conspicuously by its aridity. It is John Wesley Powell's predictive nightmare of a "heritage of conflict" that drives much of the current debate over a new water ethic; there is nothing quite like scarcity to sharpen the senses and intensify the passions. This challenge was summarized in succinct terms in a book that was published twenty years ago:

> Meaningful and enduring change in western water law and policy must be rooted in a set of principles that reflect westerners' core values and concerns. These principles must represent a coherent vision of why people care about water—a framework defining what makes policy "good" or "bad," "right" or "wrong." They must reach a fundamental level of belief so that they can guide change and not merely respond to it. They must form a water ethic.[11]

The development of this "coherent vision" is the focus of the remainder of this chapter. First, I identify reasons why we need to change—the roots of our dilemma. Second, I focus on the basic elements of a new western water ethic—the first step, in my view, to achieving significant water reform.

Three Problems

Many species, not just humans, like to draw boundaries. It is essential to the attainment of exclusivity—the ability to keep others out and reserve for ourselves, or our group, access to essential resources. Chimpanzees do it, bull elk do it, every tribe throughout history has done it, and it is at the very heart of the notion of a modern nation-state. Good fences do indeed make good neighbors, if only because a trespassing interloper can be easily identified and dispatched in short order. This is the essence of private property and is essential to modern notions of value, property, and wealth. But it can be overdone, especially when it comes to natural resources. We have divvied up the earth to a remarkable degree with political boundaries that are largely arbitrary geographically and ecologically. Balkanization is now planet-wide. The United States, for example, has 90,056 governing jurisdictions, according to the U.S. Census. And to ensure that every square inch of America is, well, square, the entire nation was divided into townships and sections. This made at least some sense in the humid East, but in the West this "parceling of the public lands," as John Wesley Powell called it, ignored the realities of aridity and vertical relief.[12]

The political balkanization of the United States is especially severe in the American West, which has a large and multifaceted federal presence and numerous Indian reservations layered over state and local jurisdictions. Utah has so many multiple layers of jurisdictions that a map of the state with color-coded land management areas is a maze of splotchy colors.[13] Federal lands in the state are managed by multiple agencies, including the Bureau of Land Management, the National Park Service, the U.S. Fish and Wildlife Service, the Defense Department, and the Forest Service. Five Indian tribes have reservations that cover about 4 percent of the state.[14] The State of Utah also owns many different kinds of lands, including state parks, school trust lands, game preserves, areas associated with universities, and sovereign lands. Another layer is added by county and municipal holdings; the state has 622 local governments, including 307 special districts, many of which deal with water.[15] In addition to these myriad layers of public governance are literally tens of thousands of parcels of private land.

Utah has also divvied up the water. The state has a complex system of private water holdings, based on the Prior Appropriation Doctrine. It has dams and other water projects built by the federal Bureau of Reclamation and the U.S. Army Corps of Engineers as well as multiple state water projects, innumerable county and local water projects, projects built by water districts, and countless private dams and diversions. A map of the state that depicts water jurisdictions is so complicated that it is virtually impossible to portray the various jurisdictions without sophisticated GIS applications.

There is nothing inherently wrong with this system of apportioning land and water, but it does present a basic governing challenge. Natural resources—indeed the planet itself—exist as a system; all the parts are connected. Fragmenting land and water into multiple layers of jurisdiction and ownership may make sense from a political standpoint, but it simply does not meld with the reality of an ecosystem. River basins are complex networks of streams, rivulets, aquifers, and mainstem channels. Watersheds encompass the land, the flora and fauna (including us), and all the complexity of nature. A change in one element of the system provokes a change in many others: "altering the fluxes of water in the landscape affects the functioning of a whole hydrogeomorphological system."[16]

The only sensible way to study, understand, and manage water is from an ecological systems perspective. Roughly paraphrasing Aldo Leopold's land ethic, a water ethic is right when it tends to preserve the integrity, stability, and beauty of the whole watershed system.[17] This is largely a corollary to modern concepts of environmental ethics, where "people can see the whole commonwealth of a human society set in its ecosystems."[18] Other scholars have used the term "ecohydrosolidarity" to describe the close, interconnected relationship between the natural world and all water users.[19] These notions are given priority in the concept known as "integrated water resource management" (IWRM), defined as "a process which promotes the coordinated development and management of water, land and related resources, in order to maximize the resultant economic and social welfare in an equitable manner without compromising the sustainability of vital ecosystems."[20]

The hydrological system binds us together; our myriad political boundaries sunder that relationship. The systemic approach to water is profoundly at odds with current management policy. When water is managed piecemeal, we lose control over these systemic impacts; they occur outside the management regime and become, in economic jargon, externalities. This inevitably harms others. Sandra Postel uses a bodily metaphor to explain this: "To manage water as if it were separate and apart from us is like cutting off the flow of blood to one part of the body in order to send it to another—the living entity suffers."[21]

A second problem is derivative of the first. In modern water law and policy the owner of each little piece of the water system can do with it what he or she pleases, with little or no regard to its effect on society as a whole or people living in other political jurisdictions.[22] Western water law contains no provisions that generally protect commonly held resources such as rivers or public lands or the larger biotic community—or, for that matter, people in other states. The only limitations on private water rights are those provided by specific, targeted legislation such as the Safe Drinking

Water Act or the Endangered Species Act. And states, as water rights holders, have even greater liberty to use water without regard to the impact on other states. The prevailing concept, to put it in colloquial terms is "I've got mine, the heck with you."

For example, a farmer in Wyoming can decide to flood-irrigate an additional field of alfalfa (assuming he or she has a state permit) without worrying about the impact of these actions on water users in Los Angeles or recreationists at Lake Powell or migratory bird species in the Colorado River delta. The City of Las Vegas can decide to build a pipeline to the Great Basin and mine groundwater from that area without considering the long-term impacts that it will have on neighboring states or Indian reservations. The city simply proclaims: it's our water. That is true, but it leaves open the question of whether such an attitude is ethical.

The third problem concerns increasing scarcity. The arid regions of the West have seen dramatic population growth in recent years—without a commensurate increase in rainfall or river flow. In Utah, for example, the population has doubled since 1980.[23] The current system of water law worked quite well when the region was sparsely populated, but it conspicuously fails to deal with the increased need for sharing, collaboration, and region-wide coordination. The stark reality that rivers are systems is much more obvious when there is not enough water for everyone. Two examples illustrate this problem.

The Prior Appropriation Doctrine—the basis of water rights in Utah and other arid states—rewards water to the earliest diverters, mainly farmers and ranchers. Today less than 2 percent of the population engages in farming or ranching but still has possession of approximately 80 percent of the water in arid states.[24] To exacerbate the problem, the Prior Appropriation Doctrine penalizes farmers for employing water-saving conservation measures; their "reward" for using water more carefully and efficiently is the loss of that part of their water right that is conserved. And the notion of leaving water in a stream to meet the needs of public lands, habitat, recreation, and other ecosystem services is wholly at odds with the doctrine.

A second example is the "law of the river" regarding the Colorado River. The 1922 Colorado Compact arbitrarily divided the basin into upper and lower regions and then awarded each an absolute amount of water (rather than a percentage of flows).[25] Following the tradition of carving up land and water, the law of the river gives each state "its" water, even though the use of that water has a direct impact on the other states. The results are predictable; the river is egregiously oversubscribed, and each state claims the right to divert its full share, regardless of the impact on others.[26] Eventually scarcity will force a new water regime to replace the old one.[27]

These three problems clearly illustrate the need for a new water ethic. Westerners can respond to this need in three contrasting ways. First, they can hire lawyers to file court cases, pay lobbyists to convince lawmakers to resist change, and make large campaign donations to any politician who promises to protect the status quo. Second, they can grudgingly recognize a changing reality and accept a minimal amount of adjustment out of necessity. A third approach is to embrace a new water ethic that profoundly changes our relationship with water and how we use it. This new ethic, tailored to fit the needs of an arid West, is the subject of the remainder of this chapter.

A New Water Ethic

The new water ethic is based on long-held general principles. Over 225 years ago Immanuel Kant developed the concept of the "categorical imperative," which states that we have a moral obligation to treat others well (even those downstream): "So act as to treat humanity, whether in your own person or in that of any other, in every case as an end in itself, never only as means."[28] In other words, people are not just objects; rather, we have an obligation to consider them in all our actions. The new water ethic goes well beyond Kant, but this quotation makes it clear that the "new" ethic has deep roots.

In 1997 the first World Water Forum made an effort to define the concept of a modern water ethic; this effort was continued at subsequent forums, the most recent held in 2012.[29] The concept was further developed by the World Commission on Ethics and Scientific Knowledge and Technology in 2004 in its report "Best Ethical Practice in Water Use."[30] Most recently the Water Ethics Network has provided an ongoing forum for a discussion of how stakeholders can "discover, support, or implement water management solutions that support the long-term interests of society and the environment."[31] For this chapter I have borrowed heavily from previous works to develop a water ethic that is tailored to the needs of the arid American West. It consists of five components.

Valuing Water

Water is a precious resource—everyone agrees with that. Yet we price water as though it were limitless and unimportant. To ensure that water is underpriced, the federal government subsidizes not only the water but massive water projects, water delivery systems, the energy used to transport water, and the resulting wastewater treatment. Western states add even more layers of subsidies and hide the cost of water with property taxes and sales taxes. If water is so precious, why do we pay so little for it? In the United States most water is used not for essential personal consumption but for commercial application. In Utah human consumption accounts for only about 2 percent of water use.[32] Given that water is a cost of business, we should charge market price for commodified water (water used as a commodity by private water users). This concept was recognized in a pivotal international conference on water held in Dublin in 1992: "Water has an economic value in all its competing uses and should be recognized as an economic good."[33] Noted water authority Robert Glennon writes: "Water is a basic commodity for which there is no substitute.... [The] redefinition of the role of water should involve pricing water in a way that reflects its importance, its uniqueness, and its finite limits."[34]

Foremost among the factors affecting the valuation of water is the notion of opportunity costs related to ecosystem services.[35] Opportunity costs are uses that are forfeited when a resource is allocated to another use. Given that water is a finite resource, nearly all water allocations are zero-sum, meaning that one use of water results in a commensurate decrease in water available to another use. For example, if a stream is diverted to water hayfields, it can no longer function effectively as a habitat for fish and wildlife. Any realistic cost-benefit analysis of such a scenario must compare the value of the hay crop to the value of the lost habitat. If the hay has less market value than the fishery, then such a diversion by definition is economically irrational. A realistic pricing of commodified water would make such a comparison part of our water use calculation. A new water ethic would carefully value not only water but water courses. Rivers per se have value to society—as scenic resources, habitat, recreation, and communal property and as a source of public drinking water. A true valuation of water must include its geographic, cultural, economic, and historical context. This is not an argument for privatization—water has so many intrinsic values and uses that the marketplace alone cannot possibly allocate such a resource fairly or in the long-term best interest of society. But I am arguing that when water is used for commercial purposes—and sold to private users—the price of the water should reflect its market value. An exception to this policy is outlined in the next section.

Water as a Human Right

A prevailing theme in the water literature is the concept that water is a human right: all decisions regarding water "must be situated in relation to deliberations on ethics, fairness, temperance, and justice."[36] This is especially relevant to those who cannot afford to buy clean, safe drinking water on the open market. As geographer Adrian Armstrong has pointed out, "in allocating water, we are effectively allocating life."[37] The World Commission on the Ethics of Scientific Knowledge and Technology recognized this crucial link between water and basic human needs in its fundamental principles: "there is no life without water and those to whom it is denied are denied life."[38]

A fundamental principle of the new water ethic must be the provision of safe drinking water to all people, regardless of income level.[39] Most people in developed countries can afford to pay their water bill, especially if they make an effort to conserve water and use it efficiently. And most of the water diverted in developed countries is for commercial application or watering turf grass. The concept of a human right to water does not apply to these additional uses but rather is aimed at guaranteeing to everyone sufficient drinking water to sustain health and well-being.

There are two notable solutions to this. One is to subsidize drinking water for the poor who otherwise could not afford it. An example comes from Chile. That country went through a period of privatizing its water sources; as a result, a large segment of the population could no longer afford drinking water. The government responded by issuing "water stamps," a program somewhat similar to the food stamp program in the United States.[40] A second method is to have a graduated price index for water based on the quantity used, known as increasing block rates. Residential users pay a very modest amount for the initial block, which is enough water for family use—but not enough to water an acre of bluegrass. As the usage increases, so does the price.[41] This makes what I call "necessity water" very cheap but makes wasteful water use very expensive. The idea of a baseline need for water can be found in the European Charter on Water Resources, which recognizes a special obligation to provide water for "vital human needs."[42] Water policy, and water pricing, must distinguish between these vital human needs and the uses of water that may be appealing but are not worthy of status as a human right. This is an important distinction in a state where people routinely overwater turf grass lawns—and the adjoining sidewalk—and pay a pittance for the water.

The concept of a guaranteed human right to water has important implications for the Prior Appropriation Doctrine. As the authors of the *Headwaters* study point out: "Water uses that promote only individual wealth and disregard broader interests of the public may not be truly beneficial. Accordingly, prior appropriation only allowed private uses because it presumed they benefitted the community. A hard look at water policy should seek distributional fairness."[43]

A second aspect of this argument is the assurance that water is safe and free of dangerous contaminants, regardless of the income level of the consumer. Most of the 2 billion people on the planet who do not have access to clean, safe water are poor and live in developing countries.[44] But even well-heeled people in developed countries confront difficult water quality problems. In the United States we face a multitude of threats from various contaminants, including mercury, chemical pesticides, fertilizers, and herbicides, perchlorate from rocket fuel, sewage discharge, lead, toxic runoff from mines, liquefied manure from concentrated animal feedlots, and radioactive pollutants.[45]

The new water ethic must address both water justice and water quality. The provision of clean "necessity water" to everyone, regardless of income or place of residence, is the only fair water policy. We cannot have justice without water justice.

No Harm

Every one of us, regardless of how "green" we live our life and how small our ecological footprint, has an impact on the planet. We are all consumers, we all produce waste, and we all take up space. Thus we all have an ethical responsibility to consider how our actions affect others (see the quotation from Kant above). In international law this precept is expressed as the "no harm" rule.[46] This means that "every state has a right to use the water in its territory on condition that this does not harm the interests of the other states."[47] Like other elements of the new water ethic, this concept is not new. John Wesley Powell recognized the need for cooperative control of water to prevent any single irrigator from working great harm to the others. Wallace Stegner described Powell's thinking: "Settlers should now be limited in their anarchic personal rights and brought up sharp against a thing that until now few had bothered to consider: the common interest.... The justification was the abiding aridity of the West.... Mutuality was a condition of survival."[48] A more recent formulation of the same idea is provided by Carolyn Merchant, who proposed the idea of a "partnership ethic...grounded in notions of relation and mutual obligation."[49]

The notion that we consider how we impact our neighbors is related to the modern U.S. water law concepts of equitable apportionment and the public trust doctrine.[50] But such considerations seldom find their way into the harshly competitive world of western water politics. A new water ethic would imbue all of us with a heightened sense of community writ large; each of us has a responsibility to consider how our actions affect others, including other species. This should be true not only for individuals but for corporations, states, localities, water districts, and cities. As water becomes scarcer, it will test our ability as a society to be considerate of others. Our new water ethic should be our guide.

Public Participation

One of the more enduring concepts in the water literature is the idea of a "hydraulic civilization," first popularized by the German historian Karl Wittfogel. The essence of this idea is that large-scale irrigation inevitably concentrates power and wealth in the hands of a despotic few.[51] Donald Worster applied the idea, with adroit modifications, to the American West, labeled it the capitalist state mode, and warned: "Democracy cannot survive where technical expertise, accumulated capital, and their combination is allowed to take command."[52]

Without a doubt, water and power tend to flow together, especially in arid regions where scarcity dictates winners and losers. The water-power nexus has been exacerbated by the policies of the U.S. government, which has handsomely subsidized some water users while ignoring others. Water projects allocate not just water but power, wealth, prestige, and security; these projects—hundreds of them—have concentrated much sought-after resources in the hands of a few, despite early proclamations that irrigation and hydropower would serve the common people.[53]

This concentration of benefits and power inevitably led to protest and calls for reform. The *Headwaters* authors contend: "There is a deeply felt sense today that water decisions, many made years ago, reflect only a narrow range of interests and concerns."[54] The past twenty years have seen a remarkable effort to democratize water policy decisionmaking. As a result one of the central tenets of the new water ethic is widespread stakeholder participation in all aspects of water policy. Gary Chamberlain calls this the rise of "water democracies."[55] Peter Brown and Jeremy Schmidt refer to it as "environmental pragmatism," which values "pluralism, deliberative dialogue, and experimental environmental policies."[56] The full implications of this new standard of citizen involvement were described by David Feldman and Helen Ingram: "Democratic decision-making for water—as for anything else—requires more than freely-contested, majoritarian-based elections: it implies representation, and direct participation of affected groups in decisions, achievement-based values in making policy choices, and decentralized, open decision making which permits wide policy debate."[57]

Achieving such a level of citizen input will require changes in law. Ana Palacio of the World Bank, speaking at a Water Forum in 2006, proposed an "ethics code of water" that would require, among other things, "a genuine participatory approach to planning and management of the resource, which would involve users, civil society organizations, and government organizations. The ethics code would, in essence, make water everybody's business."[58] A new approach to ethics cannot take place without widespread stakeholder participation. Democracy, while an imperfect instrument, is still morally superior to the alternatives. A new ethic, to be valid, must be an ethic of the people, by the people, for the people.

Stewardship

Stewardship is a modern word for an ancient practice. American Indians understood it innately. Henry David Thoreau probably never uttered the word "stewardship" but spent his life advocating for it and helped set in motion a tradition of thought that culminated in modern concepts of stewardship and sustainability. There is a long tradition in many of the world's great religions to hold water and rivers sacred.[59] Patrick Dobel sees a strong commitment to stewardship in the Judeo-Christian tradition:

> The parable of the talents makes it abundantly clear that we who are entrusted with His property will be called to account for our obligation to improve the earth. The stewardship imperative assumes that the moral and ecological constraints are respected, and it adds the obligation to distribute the benefits justly....True stewardship requires both respect for the trusteeship and covenanted imperatives and an active effort to improve the land for the future.[60]

This long tradition of caring for the planet is reflected in modern notions of a water ethic. Maude Barlow and Tony Clarke argue that "[a]t the heart of any new water ethic must be a renewal of our ties with the natural world and a reverence for water's sacred place in it."[61] Feldman and Ingram note: "Stewardship ethics are based on the premise that we are obliged to care for creation and to concern ourselves with what anyone—regardless of her or his generation—must do in order to ensure that creation is sustained."[62] The World Commission on Ethics and Scientific Knowledge and Technology, in its fundamental principles, included stewardship "which respects wise use of water."[63]

At the heart of the notion of stewardship is the belief that we all have a responsibility to care for the natural world, including all the species that depend on it for survival. This sense of responsibility is not limited only to the problems that we have caused as individuals. Because it is an ecosystem, we are all affected by negative impacts either directly or indirectly, albeit unequally. This is embodied in the moral precept that an injustice to one is an injustice to all. Thus even though I did not, for example, pollute the Duwamish River in Washington (parts of it are a toxic waste site), I still have a responsibility to assist in returning it to a state of health. When we help each other, we help ourselves; when we save the planet, we save ourselves. A new water ethic must embrace this principle.

Clearly there is some tension among these five components of a new ethic. I do not think that diminishes their validity, however, but rather points out the complexity and importance of water in every aspect of our lives; a lot of balancing and many difficult choices are embodied in this ethic. How judiciously and fairly we make such choices will test our commitment to achieving justice and sustainability for the "water planet" and all that depends on it for life.

Conclusion: Lessons for Utah

What are the implications of a new water ethic for Utah? How can we provide for our families while fulfilling the admittedly ambitious goals of this ethic? I see five ways in which the new water ethic can guide us as we move into a new era of water policy that preserves what we have, meets our needs, and yet leaves the planet in the same or better condition for future generations.

First, the new ethic requires that a sense of responsibility be attached to every declaration of a right. When my son was growing up, we had a household rule: you mess it up, you clean it up. In more formal terms, this is the "polluter pays" legal concept.[64] This emphasis on responsibility is evident in the United Nations' recent report on the world's water: *Water: A Shared Responsibility*.[65]

We need to strengthen the connection between those who damage our waterways and those who pay for remediation, clean-up, and restoration. The economic concept of externalities is central to this ethic. An externality is when an entity—an individual, a corporation, a government—figures out a way to lower its costs by passing them off to another party (including future generations or nonhuman species). For example, a factory can lower its production costs by piping its waste to the nearest stream or river. The responsibility for dealing with that waste is then passed on to everyone who lives downstream. That example is easy to visualize as an immoral act, but the same principle should be extended to water rights holders, be they individuals, cities, or states. Water users who dewater a stream or cause extensive groundwater depletion or increase the salinity of a river should be held accountable for the externalities.

Second, a new ethic requires us to be adaptable; we must change with conditions or risk becoming aliens in our own environment. Policies that were fair in one era may not be fair under current conditions. In Utah the economy has changed in fundamental ways, the population has grown dramatically, and climate change is altering our notion of "normal" precipitation and temperature. Water allocation decisions will increasingly become zero-sum, with no "surplus" to dole out to expectant petitioners. The rules for making zero-sum decisions must be different than those we relied upon when we had plentiful unallocated resources. Today every proposed increase in supply means taking water from someone or something else. Under our archaic water laws some water is still "unallocated," but in ecosystem terms every river basin in Utah is overallocated.

Third, stewardship means limits. The ecosystem has a carrying capacity; if we exceed that, we harm ourselves and future generations. The realization that we cannot grow infinitely in population, in economic terms, or in the size of our appetites is unpopular. But part of our ethic of stewardship means facing these difficult questions. What would life in Utah be like with 10 million people? How about 2 billion? Such stark numbers should expand our ethical consciousness to include the relationship between very personal decisions and the well-being of all living things—including our own children. This is not impossible; other societies have learned to live within their limits, using a combination of social/cultural change, self-imposed restraints, and government-imposed limits.[66] The new water ethic asks us to face this reality.

Fourth, Albert Einstein reputedly said, "Problems cannot be solved within the mindset that created them."[67] This points to the need for new institutions, new laws, and new approaches to managing water. We do not have a water crisis in Utah, we have a water *management* crisis. We have a crisis of innovation and a lack of vision. There is a sufficient supply of water to meet our current needs and the ecosystem that keeps us alive—if we use water wisely and efficiently. And there is sufficient water for modest, wise growth. To adjust to this new reality, we need new water institutions. We do not have to invent this wheel; innovative water management entities have been created in other states and countries. For example, France created regional "water parliaments."[68] The Global Water Partnership created innovative ways to coordinate water resource planning across multiple political jurisdictions and numerous stakeholders.[69] New systems-based water institutions are central to integrated water resource management (IWRM).[70]

At the heart of these new institutions must be a focus on ecosystems and an effort to overcome the fragmentation described earlier. The U.N. report on best ethical practice emphasizes this: "The new paradigm reminds us that our actions affect the holistic system that is our biosphere. Ecosystems and humans alike are not functioning as mere isolated machines: they are dynamic, constantly evolving due to the interactions of their components."[71] We must begin to think of rivers as a public commons, where entire river ecosystems—not just water—are managed for the public good by open, participatory water institutions.[72] Fred Pearce notes that "we need a new ethos for water—an ethos based not on technical fixes but on managing the water cycle for maximum social benefit rather than narrow self-interest."[73]

The fifth point concerns how we fund water development. For more than a century the American West has relied on the federal government to pay for enormously expensive water projects. These projects are very popular among western politicians because their constituents reap the benefits without incurring the costs. But the federal government, at $17 trillion in debt, can no longer afford to be the West's water sugar daddy. The State of Utah has also heavily subsidized some water development, using state tax money to fund water projects for some people. The most recent example of this "get someone else to pay for it" mentality is an effort by the Washington County Water Conservancy District to get the people of Utah to pay for its proposed pipeline to Lake Powell.[74] Part of our new sense of responsibility must be a willingness on the part of beneficiaries to pay; if you want a water project, you should be willing to pay for it.

The new water ethic is a combination of values, practices, and policies. It is, at its most basic level, a reorientation of how we relate to water. Vandana Shiva notes that "the need now is to combine ecology with equity, and sustainability with justice."[75] Sandra Postel argues: "The essence of such an ethic is to make the protection of water ecosystems a central goal in all that we do."[76] Similarly, the *Headwaters* authors advise: "Now is the time for the whole community to step back, take stock, and infuse water decision making with an ethic—at once new and deeply traditional—that looks ultimately to individual consciences and the long-term good of our communities, economics, and water."[77] This new ethic means abandoning what Tim Miller calls the "western water myth."[78] We must replace it with an adroit combination of cutting-edge science, biocentric systems thinking, inclusive participatory public policymaking, and a commitment to justice for all in the long run. This will not be easy; Diane Raines Ward notes that "such adjustments are not for the faint of heart."[79] But it is absolutely necessary if we are to survive and prosper.

Notes

The author gratefully acknowledges the advice and assistance of Brock Bahler in writing this chapter.

1. See generally Louis Pojman, *Ethics: Discovering Right and Wrong.*
2. Edward O. Wilson, *Consilience: The Unity of Knowledge,* 48.
3. Aldo Leopold, *A Sand County Almanac,* 203.
4. Albert Schweitzer, *Civilization and Ethics,* 125.
5. Sandra Postel, *Last Oasis: Facing Water Scarcity,* 184.
6. "Whiskey is for drinking and water is for fighting" is usually attributed to Mark Twain, but I think it more likely to have been uttered by that famous sage Johnny Walker.
7. Quoted in Donald Worster, *Rivers of Empire: Water, Aridity, and the Growth of the American West,* 327.
8. Daniel McCool, *River Republic: The Fall and Rise of America's Rivers,* 22–23.
9. Gary Chamberlain, *Troubled Waters: Religion, Ethics, and the Global Water Crisis.*
10. David Feldman, *Water,* 187.
11. Sarah Bates, David Getches, Lawrence MacDonnell, and Charles Wilkinson, *Searching Out the Headwaters: Change and Rediscovery in Western Water Policy,* 178.
12. Donald Worster, *A River Running West,* 346.
13. Trust Lands Administration, "Utah Surface Land Ownership," 2006, http://gis.utah.gov/data/sgid-cadastre/land-ownership/.
14. Utah Division of Indian Affairs, 2009, http://heritage.utah.gov/utah-division-of-indian-affairs.
15. U.S. Census, "2012 Census of Governments" (Washington, D.C.: U.S. Census Bureau, September 2013), G12-CG-ISD.
16. Adrian Armstrong, "Further Ideas towards a Water Ethic," 140.
17. Leopold's famous "land ethic," from *A Sand County Almanac,* reads in part: "A thing is right when it tends to preserve the integrity, stability, and beauty of the biotic community. It is wrong when it tends otherwise" (224–25).
18. Holmes Rolston III, *Environmental Ethics,* 246.
19. Malin Falkenmark and Carl Folke, "Ecohydrosolidarity: A New Ethics for Stewardship of Value Adding Rainfall," in *Water Ethics: Foundational Readings for Students and Professionals,* ed. Peter Brown and Jeremy Schmidt, 247.
20. Global Water Partnership, *A Handbook for Integrated Water Resources Management in Basins,* 2009, http://www.gwptoolbox.org/images/stories/Docs/gwp_inbo%20handbook%20for%20iwrm%20in%20basins_eng.pdf.
21. Postel, *Last Oasis,* 184.

22. The concept of third-party impacts in state water law is limited to fellow rights holders.
23. U.S. Census, "Population by State," 2012, http://quickfacts.census.gov/qfd/states/49000.html.
24. In Utah farmers are placed in the same workforce category with forestry, fishing and hunting, and mining. The total workforce percentage for that entire category is 2 percent. See http://governor.utah.gov/dea/demographics.html.
25. For further details on the Colorado Compact, see chapter 10 by Annette McGivney herein.
26. U.S. Bureau of Reclamation, "Colorado River Basin Water Supply and Demand Study," 2013, http://www.usbr.gov/lc/region/programs/crbstudy.html.
27. See Edward Barbanell, *Common-Property Arrangements and Scarce Resources: Water in the American West*; James Lawrence Powell, *Dead Pool: Lake Powell, Global Warming, and the Future of Water in the West*; Jonathan Waterman, *Running Dry*.

    A glimmer of this new, more cooperative approach can be seen in the 2007 agreement among Colorado River Basin states, which developed criteria for sharing the hardships of water shortages. See "Record of Decision, Colorado River Interim Guidelines for Lower Basin Shortages and the Coordinated Operations for Lake Powell and Lake Mead," November 2007, http://www.usbr.gov/lc/region/programs/strategies/RecordofDecision.pdf.
28. Immanuel Kant (1785), "Ethics Is Based on Reason," excerpted in Earl Spurin and James Swindal, *Ethics: Contemporary Readings*, 156.
29. World Water Council, http://www.worldwatercouncil.org/.
30. United Nations, World Commission on Ethics and Scientific Knowledge and Technology, "Best Ethical Practice in Water Use," 2004, http://unesdoc.unesco.org/images/0013/001344/134430e.pdf.
31. Water Ethics Network, http://waterethics.org/mission-statement/.
32. Utah Division of Water Resources, "Municipal and Industrial Water Use in Utah," 2010, http://www.water.utah.gov/Reports/MUNICIPAL%20AND%20INDUSTRIAL%20WATER%20USE%20in%20UTAH.pdf.
33. Jeremy Schmidt, "Water Ethics and Water Management," in *Water Ethics: Foundational Readings for Students and Professionals*, ed. Peter Brown and Jeremy Schmidt, 8.
34. Robert Glennon, *Unquenchable: America's Water Crisis and What to Do about It*, 317.
35. See National Research Council, Committee on Assessing and Valuing the Services of Aquatic and Related Terrestrial Ecosystems, "Valuing Ecosystem Services: Toward Better Environmental Decision-Making, 2005"; R. Costanza et al., "The Value of the World's Ecosystem Services and Natural Capital."

36. Schmidt, "Water Ethics and Water Management," 280.
37. Armstrong, "Further Ideas towards a Water Ethic," 14.
38. United Nations, World Commission on Ethics and Scientific Knowledge and Technology, "Best Ethical Practice in Water Use," 5–6, http://unesdoc.unesco.org/images/0013/001344/134430e.pdf.
39. United Nations International Covenant on Economic, Social and Cultural Rights, 1966, http://www.unhcr.org/refworld/docid/3ae6b36c0.html; Peter Gleick, "The Human Right to Water."
40. Maude Barlow and Tony Clarke, *Blue Gold: The Fight to Stop the Corporate Theft of the World's Water*, 216–17.
41. Feldman, *Water*, 138–41.
42. Council of European Ministries, "European Charter on Water Resources," October 17, 2001, https://wcd.coe.int/ViewDoc.jsp?id=231615&Site=COE.
43. Bates et al., *Searching Out the Headwaters*, 185.
44. William Ashworth, *Nor Any Drop to Drink*, 133–205; Fred Pearce, *When the Rivers Run Dry: Water—The Defining Crisis of the Twenty-First Century*, 306; United Nations, World Commission on Ethics and Scientific Knowledge and Technology, "Best Ethical Practice in Water Use," 10.
45. McCool, *River Republic*, 190–216; Environmental Protection Agency, 2013, http://water.epa.gov/drink/contaminants/.
46. United Nations, Convention on Law of the Non-navigational Uses of International Water Courses, 1997, http://dniester.org/wp-content/uploads/2009/06/9convention-on-the-law-of-the-non-navigational-uses-of-international-watercourses-engl.pdf; Vandana Shiva, *Water Wars: Privatization, Pollution, and Profit*, 76–79.
47. Barlow and Clarke, *Blue Gold*, 219.
48. Wallace Stegner, *Beyond the Hundredth Meridian: John Wesley Powell and the Second Opening of the West*, 308.
49. Carolyn Merchant, "Fish First! The Changing Ethics of Ecosystem Management," in *Water Ethics: Foundational Readings for Students and Professionals*, ed. Peter Brown and Jeremy Schmidt, 237.
50. Dan Tarlock, "The Law of Equitable Apportionment Revisited, Updated, and Restated"; Joseph Sax, "The Public Trust Doctrine in Natural Resource Law: Effective Judicial Intervention."
51. Karl Wittfogel, *Oriental Despotism: A Study of Total Power*.
52. Worster, *Rivers of Empire*, 57.
53. Donald Pisani, *Water and American Government*; Daniel McCool, *Command of the Waters: Iron Triangles, Federal Water Development, and Indian Water*; Mark Reisner, *Cadillac Desert: The American West and Its Disappearing Water*.
54. Bates et al., *Searching Out the Headwaters*.
55. Chamberlain, *Troubled Waters*.

56. Peter Brown and Jeremy Schmidt, eds., *Water Ethics: Foundational Readings for Students and Professionals,* 143.
57. David Feldman and Helen Ingram, "Multiple Ways of Knowing Water Resources: Enhancing the Status of Water Ethics," 2.
58. Ana Palacio, "Keynote Address," in *Water Ethics,* ed. Ramón Llamas, Luis Martinez-Cortina, and Aditi Mukherji.
59. Shiva, *Water Wars,* 131–39.
60. Patrick Dobel, "The Judeo-Christian Stewardship Attitude to Nature," 906.
61. Barlow and Clarke, *Blue Gold,* 211.
62. Feldman and Ingram, "Multiple Ways of Knowing Water Resources," 14.
63. The World Commission on Ethics and Scientific Knowledge and Technology, 2004: 6, http://www.unesco.org/new/en/social-and-human-sciences/themes/global-environmental-change/comest/.
64. Jean Fried, "Symposium Keynote Address. Water Governance, Management and Ethics: New Dimensions for an Old Problem."
65. *Water: A Shared Responsibility.*
66. Jared Diamond, *Collapse: How Societies Choose to Fail or Succeed;* Elinor Ostrom, *Governing the Commons;* William Ophuls and Steven Boyan, *Ecology and the Politics of Scarcity Revisited.*
67. http://www.quoteworld.org/quotes/4091.
68. Fried, "Symposium Keynote Address," 7.
69. Global Water Partnership, "Strategic Goals," 2013, http://www.gwp.org/en/Our-approach/Strategic-goals/.
70. Ian Calder, *Blue Revolution.*
71. United Nations, World Commission on Ethics and Scientific Knowledge and Technology, "Best Ethical Practice in Water Use," 11.
72. McCool, *River Republic,* 283–304.
73. Pearce, *When the Rivers Run Dry,* 308.
74. Amy Joi O'Donoghue, "Lake Powell Pipeline Foes Rejoice in Delay." For more detail, see also chapter 6 by Zachary Frankel herein.
75. Shiva, *Water Wars,* 79.
76. Postel, *Last Oasis,* 185.
77. Bates et al., *Searching Out the Headwaters,* 198.
78. Tim Miller, "Utah Water Politics in a National Perspective," 176.
79. Diane Raines Ward, *Water Wars: Drought, Flood, Folly, and the Politics of Thirst,* 208.

Selected Bibliography

Abbey, Edward. *Beyond the Wall: Essays from the Outside.* New York: Henry Holt, 1984.

———. *Desert Solitaire: A Season in the Wilderness.* New York: Ballantine Books, 1968.

———. *Down the River.* New York: Plume, 1991.

Adams, William Y. "Ninety Years of Glen Canyon Archaeology 1869–1959: A Brief Historical Sketch and Bibliography of Archaeological Investigations from J. W. Powell to the Glen Canyon Project." *Museum of Northern Arizona Bulletin* 33, no. 2 (1960).

Adler, Robert W. "Toward Comprehensive Watershed-Based Restoration and Protection for Great Salt Lake." *Utah Law Review* 99, no. 105 (1999): 99–204.

Alcott, T. I., W. J. Steenburgh, and N. F. Laird. "Great Salt Lake–Effect Precipitation: Observed Frequency, Characteristics, and Associated Environmental Factors." *Weather and Forecasting* 27, no. 4 (2012): 954–71.

Alexander, Thomas. "Stewardship and Enterprise." *Western Historical Quarterly* 25 (Autumn 1994): 340–64.

Alley, Richard B. *Earth: The Operator's Manual.* New York: W. W. Norton & Company, 2011.

Armstrong, Adrian."Further Ideas towards a Water Ethic." *Water Alternatives* 2, no. 1 (2009): 138–47.

Arrington, Leonard J., Feramorz Y. Fox, and Dean L. May. *Building the City of God: Community and Cooperation among the Mormons.* Salt Lake City: Deseret Book Company, 1976.

Ashworth, William. *Nor Any Drop to Drink.* New York: Summit Books, 1982.

Aton, James M. *The River Knows Everything: Desolation Canyon and the Green.* Logan: Utah State University Press, 2009.

Ayers, James E. "Historic Logging Camps in the Uinta Mountains, Utah." In *Forgotten Places and Things: Archeological Perspectives on American History*, edited by Albert E. Ward. Albuquerque: Center for Anthropological Studies, 1983.

———. "Standard Timber Company Logging Camps on the Mill Creek Drainage, Uinta Mountains, Utah." *Proceedings of the Society for California*

Archeology 9 (1996): 179–82. http://www.scahome.org/publications/proceedings/Proceedings.09Title%20Contents.pdf.

Barbanell, Edward. *Common-Property Arrangements and Scarce Resources: Water in the American West.* New York: Praeger, 2001.

Barlow, Maude, and Tony Clarke. *Blue Gold: The Fight to Stop the Corporate Theft of the World's Water*. New York: W. W. Norton, 2002.

Barnett, Tim, and David Pierce. "When Will Lake Mead Go Dry?" *Water Resources Research* 44, no. 3 (March 29, 2008).

Bates, Sarah, David Getches, Lawrence MacDonnell, and Charles Wilkinson. *Searching Out the Headwaters: Change and Rediscovery in Western Water Policy.* Washington, DC: Island Press, 1993.

Bear River Commission. *Bear River Compact.* Bountiful, Utah: Bear River Commission, 1958.

Bear River Development. Salt Lake City: State of Utah, Department of Natural Resources, Division of Water Resources. 2000.

Bedford, D. P. "The Great Salt Lake: America's Aral Sea?" *Environment: Science and Policy for Sustainable Development* 51, no. 5 (2009): 8–21.

———. "Utah's Great Salt Lake: A Complex Environmental-Societal System." *Geographical Review* 95, no. 1 (2005): 73–96.

Bedford, D. P., and A. Douglass. "Changing Properties of Snowpack in the Great Salt Lake Basin, Western United States, from a 26-Year SNOTEL Record." *Professional Geographer* 60, no. 3 (2008): 374–86.

Blumm, Michael C. "Public Property and the Democratization of Western Water Law: A Modern View of the Public Trust Doctrine." *Environmental Law* 19 (1989): 573–604.

Brower, David. *For Earth's Sake: The Life and Times of David Brower.* Salt Lake City: Peregrine Smith Books, 1990.

Brown, Peter, and Jeremy Schmidt, eds. *Water Ethics: Foundational Readings for Students and Professionals.* Washington, DC: Island Press, 2010.

Bryner, Gary. "Theology and Ecology: Religious Belief and Environmental Stewardship." *BYU Studies* 49, no. 3 (Summer 2010): 21–45.

Cadillac Desert: An American Nile, Water and the Transformation of Nature. (Program 2). Directed by Jon Else and Linda Harrar. New York: Homevision, 1997. VHS, 60 minutes.

Calder, Ian. *Blue Revolution.* 2nd ed. Oxford, UK: Routledge, 2005.

Carter, D. Robert. *Utah Lake: Legacy.* Provo: June sucker Recovery Implementation Program, 2002. http://utahlakecommission.org/Utah_Lake_Legacy.pdf.

Chamberlain, Gary. *Troubled Waters: Religion, Ethics, and the Global Water Crisis.* New York: Rowman & Littlefield, 2008.

Clewell, Andre F., and James Aronson. *Ecological Restoration: Principles, Values, and Structure of an Emerging Profession*. Washington, DC: Island Press, 2007.

Colton, L. J. "Early Day Timber Cutting along the Upper Bear River." *Utah Historical Quarterly* 35, no. 3 (Summer 1967): 202–8.

Conover, Beth, ed. *How the West Was Warmed: Responding to Climate Change in the Rockies*. Boulder: Fulcrum Press, 2009.

Costanza, R., et al. "The Value of the World's Ecosystem Services and Natural Capital." *Nature* 387 (May 15, 1997): 253–60.

Crampton, C. Gregory. *Ghosts of Glen Canyon: History beneath Lake Powell*. Rev. ed. Salt Lake City: Bonneville Books/University of Utah Press, 2009. First published 1986 by Publishers Place.

Crimmel, Hal. *Dinosaur: Four Seasons on the Green and Yampa Rivers*. Tucson: University of Arizona Press, 2007.

Cui, Yue, Ed Mahoney, and Teresa Herbowicz, eds. *Economic Benefits to Local Communities from National Park Visitation, 2011*. Fort Collins, CO: U.S. Department of the Interior, National Park Service, 2013. http://www.nature.nps.gov/socialscience/docs/NPSSystemEstimates2011.pdf.

DeBuys, William. *Great Aridness: Climate Change and the Future of the American Southwest*. New York: Oxford University Press, 2011.

Denton, Craig. *Bear River: Last Chance to Change Course*. Logan: Utah State University Press, 2007.

Diamond, Jared. *Collapse: How Societies Choose to Fail or Succeed*. Rev. ed. New York: Penguin Books, 2011.

Dobb, Edwin. "The New Oil Landscape." *National Geographic* 223, no. 3 (March 2013): 28–59.

Dobel, Patrick. "The Judeo-Christian Stewardship Attitude to Nature." *Christian Century*, October 12, 1977.

"Early Days in Ogden." *Deseret Weekly*, February 23, 1895.

Farmer, Jared. *On Zion's Mount: Mormons, Indians, and the American Landscape*. Cambridge, MA: Harvard University Press, 2008.

Feldman, David. L. *Water*. Malden, MA: Polity Press, 2012.

Feldman, David, and Helen Ingram, "Multiple Ways of Knowing Water Resources: Enhancing the Status of Water Ethics." *Santa Clara Journal of International Law* 7, no. 1 (2009). http://digitalcommons.law.scu.edu/scujil/vol7/iss1/1/.

"Final Great Salt Lake Comprehensive Management Plan and Record of Decision." State of Utah. Department of Natural Resources. Division of Forestry, Fire, and State Lands. March 2013. http://www.ffsl.utah.gov/sovlands/greatsaltlake/2010Plan/OnlineGSL-CMPandROD-March2013.pdf.

Fishman, Charles. *The Big Thirst: The Secret Life and Turbulent Future of Water.* New York: Free Press, 2012.

Fleck, Richard F., ed. *A Colorado River Reader.* Salt Lake City: University of Utah Press, 2000.

Fradkin, Philip L. *A River No More: The Colorado River and the West.* Tucson: University of Arizona Press, 1984.

Frank, Matthew. "Montana's Stream Access Law Stays Strong." *High Country News,* July 25, 2011. http://www.hcn.org/issues/43.12/montanas-stream-access-law-stays-strong.

Fried, Jean. "Symposium Keynote Address, Water Governance, Management and Ethics: New Dimensions for an Old Problem." *Santa Clara Journal of International Law* 6, no. 1 (2008): 1–14.

Frye, Bradford J. *From Barrier to Crossroads: An Administrative History of Capitol Reef National Park, Utah.* USDI National Park Service, Intermountain Region. Cultural Resources Selections. Denver, Colorado, 1998. http://www.nps.gov/history/history/online_books/care/adhi/adhi.htm.

Gangopadhyay, S., and T. Pruitt. "Hydrologic Projections for the Western United States." *Eos, Transactions, American Geophysical Union* 92, no. 48 (2011): 441–42.

Gasland. Directed by Josh Fox. New York: New Video Group, 2010. DVD, 107 minutes.

Gasland II. Directed by Josh Fox. New York: New Video Group, 2013. DVD, 125 minutes.

Gessner, David. "How Big Oil Seduced and Dumped This Utah Town." *Mother Jones.* March 19, 2013. http://www.motherjones.com/environment/2013/03/how-vernal-utah-learned-love-big-oil.

Gillespie, J., S. T. Nelson, A. L. Mayo, and D. G. Tingey. "Why Conceptual Groundwater Flow Models Matter: A Trans-Boundary Example from the Arid Great Basin, Western USA." *Hydrogeology Journal* 20 (2012): 1133–47.

Gillies, R. R., S-Y Wang, and M. R. Booth. "Observational and Synoptic Analyses of the Winter Precipitation Regime Change Over Utah." *Journal of Climate* 25, no. 13 (2012): 4679–98.

Gleick, Peter. "The Human Right to Water." *Water Policy* 1, no. 5 (1998): 487–503.

Glennon, Robert. *Unquenchable: America's Water Crisis and What to Do about It.* Washington, DC: Island Press, 2009.

Glick, Daniel. "A Dry Red Season: Drought Drains Lake Powell, Uncovering the Glory of Glen Canyon." *National Geographic* 209, no. 4 (April 2006): 64–81.

Gloss, Steven P., Jeffrey Lovich, and Theodore Melis, eds. *The State of the Colorado*

River Ecosystem in Grand Canyon: A Report of the Grand Canyon Monitoring and Research Center, 1991–2004. Reston, VA: U.S. Geological Survey, 2005.

Gowans, Matthew, and Philip Cafaro. "A Latter-day Saint Environmental Ethic." *Environmental Ethics* 25 (2003): 375–94.

Gruver, Mead, and Ben Neary. "EPA Won't Finalize Wyoming Fracking-Pollution Study." Associated Press, June 20, 2013. http://bigstory.ap.org/article/epa-wont-finalize-wyo-fracking-pollution-study.

Gwynn, J. W., ed. *Great Salt Lake: An Overview of Change*. Special Publication of the Utah Department of Natural Resources. Salt Lake City: Department of Natural Resources, 2002.

Haines, Leslie. "Uinta Basin." Excerpted from *Oil and Gas Investor*. May 2012. http://www.newfield.com/assets/pdf/Uinta.reprint.5-12.pdf.

Hamlet, A. F., P. W. Mote, M. P. Clark, and D. P. Lettenmaier. "Effects of Temperature and Precipitation Variability on Snowpack Trends in the Western United States." *Journal of Climate* 18, no. 21 (2005): 4545–61.

Handley, George. "The Environmental Ethics of Mormon Belief." *BYU Studies* 40, no. 2 (Summer 2001): 187–211.

———. *Home Waters: A Year of Recompenses on the Provo River*. Salt Lake City: University of Utah Press, 2010.

———. "The Poetics of the Restoration." *BYU Studies* 49, no. 5 (Autumn 2010): 45–72.

Handley, George, Terry Ball, and Steven Peck, eds. *Stewardship and the Creation: LDS Perspectives on the Environment*. Provo: Religious Studies Center, 2006.

Henetz, Patty. "Conservation Report Card: Utah Trying to Cut Use, But Still a Top Water Guzzler." *Salt Lake Tribune*. November 9, 2009. http://www.sltrib.com/news/ci_13750549.

Hesse, Hermann. *Siddhartha*. Translated by Hilda Rosner. Project Gutenburg, Kindle, 2008. First published 1922 by S. Fischer Verlag.

Hoffman, Edward, ed. *The Wisdom of Carl Jung*. New York: Citadel, 2003.

Hostetler, S. W., F. Giorgi, G. T. Bates, and P. J. Bartlein. "Lake-Atmosphere Feedbacks Associated with Paleolakes Bonneville and Lahontan." *Science* 263, no. 5147 (1994): 665–68.

Hundley, Norris, Jr. *Water and the West: The Colorado River Compact and the Politics of Water in the American West*. 2nd ed. Berkeley: University of California Press, 2009.

Hylton, Wil S. "Broken Heartland." *Harper's* (July 2012): 25–35.

Jibson, Wallace N. *History of the Bear River Compact*. Logan, UT: Bear River Compact Commission, 1990.

Karl, T., J. M. Melillo, and T. C. Peterson, eds. *Global Climate Change Impacts in the United States*. Cambridge: Cambridge University Press, 2009.

King, D. T., and D. W. Anderson. "Recent Population Status of the American White Pelican: A Continental Perspective." *Waterbirds: The International Journal of Waterbird Biology* 28, Special Publication 1 (2005): 48–54.

Kolbert, Elizabeth. *Field Notes from a Catastrophe: Man, Nature and Climate Change*. New York: Bloomsbury USA, 2006.

"Lake Powell Pipeline Could Quadruple Water Costs, U. Economists Say." KSL .com Utah. October 16, 2012. http://www.ksl.com/?nid=148&sid=22582832.

Lambert, Roy. *High Uintas Hi!* Salt Lake City: Publishers Press, 1964.

Lee, Katie. *All My Rivers Are Gone: A Journey of Discovery through Glen Canyon*. Boulder, CO: Johnson Books, 1998.

Leopold, Aldo. *A Sand County Almanac*. Oxford: Oxford University Press, 1949.

Llamas, Ramón, Luis Martinez-Cortina, and Aditi Mukherji, eds. *Water Ethics*. The Netherlands: CRC Press, 2009.

Loomis, Brandon. "Protestors Gather to Thwart Green River Nuclear Plans." *Salt Lake Tribune*, May 20, 2012. http://www.sltrib.com/sltrib/news/54147624-78/river-plant-nuclear-green.html.csp.

Lorenz, E. N. *The Essence of Chaos*. Seattle: University of Washington Press, 1993.

Lustgarten, Abrahm. "Drilling Industry Says Diesel Use Was Legal." *ProPublica*. February 8, 2011. http://wyofile.com/propublica/diesel-fracking/.

Lynas, Mark. *The God Species: How the Planet Can Survive the Age of Humans*. London: 4th Estate, 2011.

Madsen, Brigham D. *The Shoshoni Frontier and the Bear River Massacre*. Salt Lake City: University of Utah Press, 1985.

———, ed. *Exploring the Great Salt Lake: The Stansbury Expedition of 1849–50*. Salt Lake City: University of Utah Press, 1989.

Madsen, David B., and James F. O'Connell, eds. *Man and Environment in the Great Basin*. Washington, DC: Society of American Archaeology, 1982.

Maffly, Brian."Legislature Decides to Bow Out While Courts Deal with Public Access Issues." *Salt Lake Tribune*, February 19, 2013. http://m.sltrib.com/sltrib/mobile2/55858215-218/access-bill-bills-pitcher.html.csp.

Martin, Russell. *A Story That Stands Like a Dam: Glen Canyon and the Struggle for the Soul of the West*. New York: Henry Holt & Company, 1989.

McCool, Daniel. *Command of the Waters: Iron Triangles, Federal Water Development, and Indian Water*. Berkeley: University of California Press, 1987.

———. *River Republic: The Fall and Rise of America's Rivers*. New York: Columbia University Press, 2012.

———, ed. *Waters of Zion: The Politics of Water in Utah*. Salt Lake City: University of Utah Press, 1995.

McGivney, Annette. *Resurrection: Glen Canyon and a New Vision for the American West.* Seattle: Braided River, 2009.

McKibben, Bill, ed. *The Global Warming Reader: A Century of Writing about Climate Change.* New York: Penguin Books, 2012.

McPhee, John. *Encounters with the Archdruid.* New York: Farrar, Straus & Giroux, 1971.

Meloy, Ellen. *Raven's Exile: A Season on the Green River.* Tucson: University of Arizona Press, 2003.

Micklin, P. P. "Desiccation of the Aral Sea: A Water Management Disaster in the Soviet Union." *Science* 241, no. 4870 (1988): 1170–76.

Miller, Char, ed. *River Basins of the American West: A High Country News Reader.* Corvallis: Oregon State University Press, 2009.

———, ed. *Water in the 21st Century West: A High Country News Reader.* Corvallis: Oregon State University Press, 2009.

———, ed. *Water in the West: A High Country News Reader.* Corvallis: Oregon State University Press, 2000.

Miller, Tim. "Utah Water Politics in a National Perspective." In *Waters of Zion,* edited by Daniel C. McCool. Salt Lake City: University of Utah Press, 1995.

Mohammed, I. N., and D. G. Tarboton. "An Examination of the Sensitivity of Great Salt Lake to Changes in Inputs." *Water Resources Research* 48, no. 11 (2012).

———. "On the Interaction between Bathymetry and Climate in the System Dynamics and Preferred Levels of Great Salt Lake." *Water Resources Research* 47, no. 2 (2011).

Morgan, Dale L. *The Great Salt Lake.* Albuquerque: University of New Mexico Press, 1947.

Mubako, S. T., and C. L. Lant. "Agricultural Virtual Water Trade and Water Footprint of U.S. States." *Annals of the Association of American Geographers* 103, no. 2 (2013): 385–96.

Mulroy, Patricia. "Extended Interview, Desert Wars: Water and the West." KUED, September 2006. http://www.kued.org/productions/desertwars/mulroy_patricia.php.

———. Interview. KVBC Television, April 2010. http://www.ksl.com/?nid=148&sid=10387603&pid=1.

Murphy, Miriam B. *A History of Wayne County.* Utah State Historical Society/Wayne County Commission. 1999. http://utah.ptfs.com/Data/Library2/publications/dc019652.pdf.

Myers, Tom. "Loss Rates from Lake Powell and Their Impact on Management of the Colorado River." *Journal of the American Water Resources Association* 49, no. 5 (2013): 1213–24.

Nibley, Hugh. "Subduing the Earth." In *On the Timely and the Timeless*, 95–110. Provo: Religious Studies Center, 1978.

O'Donoghue, Amy Joi. "Lake Powell Pipeline Foes Rejoice in Delay." *Deseret News*, February 8, 2013. http://www.deseretnews.com/article/865572677/Lake-Powell-pipeline-foes-rejoice-in-delay.html?pg=all.

———. "Uintah Basin Ozone Problem Triggers Lawsuit against EPA." *Deseret News*, July 23, 2012. http://www.deseretnews.com/article/865559426/Uinta-Basin-ozone-problem-triggers-lawsuit-against-EPA.html?pg=all.

———. "Utah Faces Polluted Water Woe." *Deseret News*, May 18, 2013: A1.

Onton, D. J., and W. J. Steenburgh. "Diagnostic and Sensitivity Studies of the 7 December 1998 Great Salt Lake-Effect Snowstorm." *Monthly Weather Review* 129, no. 6 (2001): 1318–38.

Ophuls, William, and Steven Boyan. *Ecology and the Politics of Scarcity Revisited*. New York: W. H. Freeman, 1992.

Ostrom, Elinor. *Governing the Commons*. New York: Cambridge University Press, 1990.

Otsuka, Julie. *When the Emperor Was Divine*. New York: Anchor, 2003.

Peacock, S. "Projected Twenty-First-Century Changes in Temperature, Precipitation, and Snow Cover over North America in CCSM4." *Journal of Climate* 25, no. 13 (2012): 4405–29.

Pearce, Fred. *When the Rivers Run Dry: Water—The Defining Crisis of the Twenty-First Century*. Boston: Beacon Press, 2007.

Peterson, F. Ross, and Robert E. Parson. *Ogden City: Its Governmental Legacy—A Sesquicentennial History*. Ogden, UT: Chapelle Limited, 2001.

Petoukhov, V., and V. A. Semenov. "A Link between Reduced Barents–Kara Sea Ice and Cold Winter Extremes over Northern Continents." *Journal of Geophysical Research: Atmospheres* 115, no. D21 (2010).

Pisani, Donald. *Water and American Government*. Berkeley: University of California Press, 2002.

Pojman, Louis. *Ethics: Discovering Right and Wrong*. 3rd ed. Belmont, CA: Wadsworth, 1989.

Polson, Jim. "Frac Focus Fails as Fracking Disclosure Tool, Study Finds." *Bloomberg News*. April 23, 2013. http://www.bloomberg.com/news/2013-04-23/fracfocus-fails-as-fracking-disclosure-tool-study-finds.html.

Porter, Eliot. *The Place No One Knew: Glen Canyon on the Colorado*. 25th Anniversary Commemorative Edition. Layton, UT: Gibbs Smith, 2000. First published 1963 by Ballantine Books.

Postel, Sandra. *Last Oasis: Facing Water Scarcity*. New York: W. W. Norton, 1997.

Postel, Sandra, and Brian Richter. *Rivers for Life: Managing Water for People and Nature*. Washington, DC: Island Press, 2003.

Powell, James Lawrence. *Dead Pool: Lake Powell, Global Warming, and the Future of Water in the West.* Berkeley: University of California Press, 2009.

Powell, John Wesley. *The Exploration of the Colorado River and Its Canyons.* New York: Penguin Books, 1997. First published 1895 by the Smithsonian Institution.

Rajagopalan, Balaji, Kenneth Nowak, James Prairie, Martin Hoerling, Benjamin Harding, Joseph Barsugli, Andrea Ray, and Bradley Udall. "Water Supply Risk on the Colorado River: Can Management Mitigate?" *Water Resources Research* 45, no. 8 (2009).

Rao, Vikram. *Shale Gas: The Promise and the Peril.* Research Triangle Park, NC: RTI Press, 2012.

Reisner, Marc. *Cadillac Desert: The American West and Its Disappearing Water.* New York: Penguin, 1986.

Rich, E. E., ed. *Ogden's Snake Country Journals, 1824–26.* London: Hudson's Bay Record Society, 1979.

Richman, Jana. *The Ordinary Truth.* Torrey, UT: Torrey House Press, 2012.

Roberts, B. H. *A Comprehensive History of the Church of Jesus Christ of Latter-day Saints.* Salt Lake City: Deseret News Press, 1930.

Roberts, Richard C., and Richard W. Sadler. *A History of Weber County.* Salt Lake City: Utah State Historical Society, 1997.

Rolston, Holmes, III. *Environmental Ethics.* Philadelphia: Temple University Press, 1988.

Sax, Joseph. "The Public Trust Doctrine in Natural Resource Law: Effective Judicial Intervention." *Michigan Law Review* 68 (1970): 475–566.

Schorr, David. *The Colorado Doctrine: Water Rights, Corporations, and Distributive Justice on the American Frontier.* New Haven: Yale University Press, 2012.

Schweitzer, Albert. *Civilization and Ethics.* Translated by John Naish. London: A. & C. Black, 1929. First published 1923 by Paul Haupt.

Seager, R., M. Ting, I. Held, Y. Kushnir, J. Lu, G. Vecchi, H.-P. Huang, N. Harnik, A. Leetmaa, N.-C. Lau, C. Li, J. Velez, and N. Naik. "Model Projections of an Imminent Transition to a More Arid Climate in Southwestern North America." *Science* 316, no. 5828 (2007): 1181–84.

Shiva, Vandana. *Water Wars: Privatization, Pollution, and Profit.* Cambridge, MA: South End Press, 2002.

Sloan, Edward L., ed. *Gazeteer [sic] of Utah and Salt Lake City.* Salt Lake City: Salt Lake Herald Publishing Company, 1874.

Smart, Christopher. "Utah Group Asks Court to Rule on Stream Access Law." *Salt Lake Tribune,* September 7, 2011. http://www.sltrib.com/sltrib/news/52529408-78/public-utah-coalition-court.html.csp.

Smith, Linda H. *A History of Morgan County.* Salt Lake City: Utah State Historical Society, 1999.

Snow, Anne, compiler. *Rainbow Views: A History of Wayne County*. 4th ed. Springville, Utah: Art City Publishing, 1985.

Solomon, Steven. *Water: The Epic Struggle for Wealth, Power, and Civilization*. New York: Harper, 2010.

Solomon, S., D. Qin, M. Manning, Z. Chen, M. Marquis, K. B. Averyt, M. Tignor, and H. L. Miller, eds. *Climate Change 2007: The Physical Science Basis*. Contribution of Working Group I to the Fourth Assessment Report of the Intergovernmental Panel on Climate Change. Cambridge: Cambridge University Press, 2007.

"Southern Nevada Water Authority Clark, Lincoln, and White Pine Counties Groundwater Development Project Conceptual Plan of Development." Southern Nevada Water Authority. March 2011.http://www.snwa.com/assets/pdf/wr_gdp_concept_plan_2011.pdf.

Spellman, Frank R. *Environmental Impacts of Hydraulic Fracturing*. Boca Raton: CRC Press, 2013.

Spurin, Earl, and James Swindal. *Ethics: Contemporary Readings*. New York: Routledge, 2004.

Stegner, Wallace. *The American West as Living Space*. Ann Arbor: University of Michigan Press, 1987.

———. *Beyond the Hundredth Meridian: John Wesley Powell and the Second Opening of the West*. New York: Penguin Books, 1953.

———. *The Sound of Mountain Water*. Lincoln: University of Nebraska Press, 1985.

Tarlock, Dan. "The Law of Equitable Apportionment Revisited, Updated, and Restated." *University of Colorado Law Review* 56 (1985): 381–412.

Trimble, Stephen. *Bargaining for Eden: The Fight for the Last Open Spaces in America*. Berkeley: University of California Press, 2008.

Utah Rivers Council. *Crossroads Utah: Utah's Climate Future*. Salt Lake City: Utah Rivers Council, 2012.

Utah State Water Plan: Bear River Basin. Salt Lake City: Utah Division of Water Resources, January 1992.

"Utah's Thirst for Water Comes with $13.7 Billion Price Tag." *Deseret News*, October 26, 2012. http://www.deseretnews.com/article/865565407/Utahs-thirst-for-water-comes-with-137-billion-price-tag.html.

Voge, Adam. "Plans for 23,000 Wells Buoy Wyoming's Oil and Gas Future." *Casper Star-Tribune*, April 1, 2013. http://trib.com/business/energy/plans-for-wells-buoy-wyoming-s-oil-and-gas-future/article_f3d3ec28-41a5-566c-bde1-3009881dbe48.html.

Wang, S-Y., R. R. Gillies, and T. Reichler. "Multidecadal Drought Cycles in the Great Basin Recorded by Great Salt Lake: Modulation from a Transition-Phase Teleconnection." *Journal of Climate* 25, no. 5 (2012): 1711–21.

Ward, Diane Raines. *Water Wars: Drought, Flood, Folly, and the Politics of Thirst.* New York: Riverhead Publishing, 2002.

Waring, Gwendolyn. *River and Dam Management: A Review of the Bureau of Reclamation's Glen Canyon Environmental Studies.* Washington, DC: National Academy Press, 1987.

Water: A Shared Responsibility. United Nations World Water Development Report 2. New York: Berghahn Books, 2006.

Waterman, Jonathan. *Running Dry*. Washington, DC: National Geographic, 2010.

Wilkinson, Charles F. *Crossing the Next Meridian: Land, Water, and the Future of the West.* Washington, DC: Island Press, 1992.

Williams, Terry Tempest. *Refuge: An Unnatural History of Family and Place.* New York: Pantheon Books, 1991.

Williams, Terry Tempest, William B. Smart, and Gibbs M. Smith, eds. *New Genesis: A Mormon Reader on Land and Community*. Layton: Gibbs Smith Publishers, 1998.

Williamson, Jeremiah. "Stream Wars: The Constitutionality of the Utah Public Waters Access Act." *University of Denver Water Law Review* 14, no. 2 (July 2011): 315–35. Also available at http://papers.ssrn.com/sol3/papers.cfm?abstract_id=1785695.

Wilson, Edward O. *Consilience: The Unity of Knowledge.* New York: Alfred A. Knopf, 1998.

———. *The Creation: An Appeal to Save Life on Earth.* New York: W. W. Norton, 2006.

Wittfogel, Karl. *Oriental Despotism: A Study of Total Power.* New York: Random House, 1957.

Woodhouse, Connie, and Jeffrey Lukas. "Multi-Century Tree-Ring Reconstructions of Colorado Streamflow for Water Resource Planning." *Climatic Change* 78 (2006): 293–315.

Woodhouse, C. A., D. M. Meko, G. M. MacDonald, D. W. Stahle, and E. R. Cook. "A 1,200-Year Perspective of 21st Century Drought in Southwestern North America." *Proceedings of the National Academy of Sciences of the United States of America* 107, no. 50 (2010): 21283–88.

Worster, Donald. *A River Running West.* New York: Oxford University Press, 2001.

———. *Rivers of Empire: Water, Aridity, and the Growth of the American West.* New York: Oxford University Press, 1985.

Yeager, K. N., W. J. Steenburgh, and T. I. Alcott. "Contributions of Lake-Effect Periods to the Cool-Season Hydroclimate of the Great Salt Lake Basin." *Journal of Applied Meteorology and Climatology* 52, no. 2 (2013): 341–62.

Young, Michael K., David Haire, and Michael A. Bozek. "The Effect and Extent of Railroad Tie Drives in Streams of Southeastern Wyoming." *Western Journal of Applied Forestry* 9, no. 4 (1994): 125–30.

Zellmer, Sandra B., and Jessica Harder. "Unbundling Property in Water." *Alabama Law Review* 59 (2007–8): 679–745.

Contributors

Daniel Bedford is a professor in the Geography Department at Weber State University. He received his BA from Oxford University in England and MA and PhD degrees from the University of Colorado in Boulder. He continues to be active in teaching and scholarship focusing on climate change and its effects on water resources, with special interests in Great Salt Lake and the public understanding of climate change.

Hal Crimmel is Rodney H. Brady Presidential Distinguished Professor of English at Weber State University and founding co-chair of WSU's Environmental Issues Committee. His publications include *Dinosaur: Four Seasons on the Green and Yampa Rivers*; *Teaching about Place: Learning from the Land*; and *Teaching in the Field: Working with Students in the Outdoor Classroom*. Crimmel directs the Master of Arts in English Program and teaches writing and literature, including field-based courses, in Montana, Colorado, and Utah.

Sara Dant is a professor of history at Weber State University, where she teaches courses on the American West and environmental history. She is the author of several prize-winning articles on western environmental politics and the co-author of the two-volume *Encyclopedia of American National Parks*. She is currently writing a book on the environmental history of the West, a chapter on Utah's environmental history since 1945, and a more detailed history of the Weber River.

Craig Denton is a professor of communication at the University of Utah and the author of *Bear River: Last Chance to Change Course* and *People of the West Desert: Finding Common Ground*. His latest photographic documentary is the website hiddenwater.org, which looks at the origins and uses of surface water on the east side of Salt Lake Valley.

Rob Dubuc is staff attorney for the Salt Lake City office of Western Resource Advocates, a nonprofit environmental law and policy organization dedicated to protecting the West's land, air, and water. In addition to protecting Great Salt Lake, his law practice focuses on guarding against impacts from tar sands and oil shale development in the Book Cliffs area, off-road vehicle damage in Utah's National Forests, and industrial air pollution along the Wasatch Front. He has a BA from the University of Maryland, an MBA from the University of Montana, and a JD from the University of Utah.

ERIC C. EWERT is a professor of geography at Weber State University and has lived in western states from Arizona to Alaska. As a lifelong and mesmerized observer of the region, his current research and writing focus on the rapid demographic and economic change occurring in the American West and the costs associated with such environmental and cultural transgressions.

ZACHARY FRANKEL is the founder and executive director of the Utah Rivers Council, which he started in 1995. He received his BS in biology at the University of Utah and has led successful grassroots campaigns to protect Utah's rivers and the ecosystems they support. He also has authored and lobbied for legislation that passed in the Utah Legislature. When he's not researching, advocating, or writing about water Frankel lives with his wife and children and their horses in Sandy, Utah.

GEORGE HANDLEY is a professor of interdisciplinary humanities at Brigham Young University and is the author of *New World Poetics: Nature and the Adamic Imagination of Whitman, Neruda, and Walcott* and an environmental memoir, *Home Waters: A Year of Recompenses on the Provo River*. He has lectured and written extensively on the links between religion, literature, and the environment.

DANIEL MCCOOL is a professor of political science, director of the Environmental and Sustainability Studies Program, and co-director of Sustainability Curriculum Development at the University of Utah. He has authored, co-authored, or edited nine books, including *River Republic, Native Waters, Command of the Waters,* and *Waters of Zion.*

ANNETTE MCGIVNEY is southwest editor for *Backpacker* magazine and a senior lecturer in journalism at Northern Arizona University in Flagstaff. She is author of *Resurrection: Glen Canyon and a New Vision for the American West* and *Leave No Trace: A Guide to the New Wilderness Etiquette*. She writes frequently about outdoor experiences and environmental issues for *Backpacker, Outside,* and *Arizona Highways.*

JANA RICHMAN is the author of a memoir, *Riding in the Shadows of Saints: A Woman's Story of Motorcycling the Mormon Trail,* and two novels, *The Last Cowgirl* and *The Ordinary Truth.* In her writing Richman explores issues that threaten to destroy the essence of the West: overpopulation, overdevelopment, rapidly dwindling water aquifers, stupidity, ignorance, arrogance, and greed. She also writes about passion, beauty, and love. Richman is currently working on a collection of essays.

STEPHEN TRIMBLE teaches writing in the University of Utah Honors College and spent a year as a Wallace Stegner Centennial Fellow at the University of Utah Tanner Humanities Center. Trimble has published more than twenty award-winning books, including *Bargaining for Eden: The Fight for the Last Open Spaces in America*; *The Sagebrush Ocean: A Natural History of the Great Basin*; and *The People: Indians of the American Southwest.* He makes his home in Salt Lake City and in the redrock country of Torrey, Utah.

Brooke Williams's conservation career spans thirty years, most recently with the Southern Utah Wilderness Alliance. He has an MBA in Sustainable Business from the Bainbridge Graduate Institute and is a freelance journalist who has published four books, including *Halflives: Reconciling Work and Wildness,* and dozens of articles. He's hard at work on a book about discovering places where the outer and inner wilderness meet. Williams, his wife (the writer Terry Tempest Williams), and two fine dogs split their time between Castle Valley, Utah, and Jackson Hole, Wyoming.

Index